COSMIC MECHANICS

COSMIC MECHANICS:

HOW WE CREATE REALITY WITH ZERO-POINT AND QUANTUM WAVES

By Dr. F. Lee Aeilts   27 July 2006

Printed by  .com

Dr. F. Lee Aeilts

COSMIC MECHANICS
Copyright 2007 by F. Lee Aeilts

All rights reserved. No part of this book may be used or reproduced in any manner whatsoever without written permission from the author, except in the case of brief quotations used with acknowledgement of the author.

Published 2007 by F. Lee. Aeilts

Dr. Aeilts discovered that our personal orbs are insidious and diabolical quantum computers that we take with us between lives.

www.cosmicmechanics.com

Printed by www.lulu.com in the United States of America

ISBN 978-0-6151-5921-8

Cover art by F. Lee Aeilts

First Edition

Address all inquiries to:
F. Lee Aeilts
1410 L. Ron Hubbard Way
Los Angeles, CA
90027

ACKNOWLEDGEMENTS:

Thanks to the following individuals for their contributions to science and philosophy which support the scientific validity of this book.

Courtney Brown	Rene Descartes	Christian Doppler	Albert Einstein
Sigmond Freud	Galileo Galilei	Graham Hancock	Werner Heisenberg
Johannes Kepler	E. Kubler-Ross	Mark McCutcheon	Raymond Moody
Issac Newton	Nikola Tesla	L. Ron Hubbard	Lynne McTaggart
Xavier Borg	Bruce H. Lipton	Milo Wolff	Ervin Lazlo

Mahatma Gandhi showed an understanding of the big picture when he wrote:

Your beliefs become your thoughts

Your thoughts become your words

Your words become your actions

Your actions become your habits

Your habits become your values

Your values become your destiny

FOREWORD

Cosmic means "of or belonging to the cosmos; having to do with the whole universe." The title of this book actually is much broader in scope than just having to do with the Physical Universe (represented here by an artists conception of a galaxy) because there are an infinite number of universes of which our Physical Universe is just the one that we are currently trapped inside.

Mechanics is the branch of Physics dealing with the action of forces on solids, liquids and gases at rest or in motion. It is the mechanical part or technique of doing something. As in Quantum Mechanics, it is how the waves work and interplay. If you don't know how the energy exchanges occur in electricity, gravity, inertia or magnetism, you don't understand these forces even if you glibly quote Newton's laws of gravity and Maxwell's formulas.

Zero-point longitudinal waves are a newly discovered kind of electromagnetic energy that is altogether different from what humans experience, e.g. radio, TV, cell phones, microwave ovens, etc. The ordinary EM waves that we have known about are called **transverse** EM waves, to distinguish them from the new **longitudinal** EM waves. These longitudinal waves do not actually exist in our "material" world, but they underlie it. They are an energetic product of the thought universe where we go between lives. They exist even in the vacuum of space. This vacuum of space we speak of exists all through everything. Even our bodies are mostly empty space between atoms and molecules. So the gateway to this seething ocean of energy can be there at every point in the universe. This seething ocean of energy is all around us and all through us.

In the 17th century, it was thought that a totally empty volume of space could be created by simply removing all matter and, in particular, all gases. That was our first concept of the vacuum. Just get rid of all the gas.

Late in the 19th century, it became apparent that the region still contained thermal radiation. But it seemed that the radiation might be eliminated by cooling. So the second concept of getting a real vacuum is to cool it down to zero temperature. Just go all the way to absolute zero. Then we've got a real vacuum, right? Well, since then, both theory and experiment have shown that **there is a non-thermal radiation in the vacuum** and that **it persists** even if the temperature could be **lowered to absolute zero**. Therefore, it was simply called the "**zero point**" radiation. Later discoveries revealed that spiritual beings like you and me have the ability to produce this zero-point radiation. It is done with our imaginations. When I ask you to get a picture of a red sports car in your mind, you do it with zero-point energy. We can direct this zero-point radiation to produce quantum energy wave structures like electrons and photons which are the building blocks of the material universe which we can describe with Newtonian Physics. This fact theoretically gives humans the godly ability to create matter out of nothing. That is the subject of this book; how we create and control material objects with zero-point and quantum waves.

Cosmic Mechanics is therefore how spiritual beings create and control universes through the emission and recording of zero-point radiation which results in quantum wave structures not sensed by humans but which exhibit the properties of charge, mass, spin and time. Vast numbers of these quantum wave structures result in our material universe which can be described by classical Newtonian physics that humans experience as solid, energetic, spacious and moving through time.

PROBLEMS WITH MODERN PHYSICS

The development of Modern Physics has "revealed" the mechanism of the origin of the Physical Universe, the origin of species, and the evolution of life. But how valid is it? Is there any truth to the "Big Bang" theory? Is a human being only a survival machine powered by chemicals and genetic coding? Is DNA the brain of a cell? Is the brain the seat of consciousness, the repository of memory and the source of our intelligence? Does nothing travel faster than the speed of light? How deep will scientists have to peer into the heart of the atomic nucleus before they will find the actual solid particles that determine its mass? So far they haven't found anything solid even with the use of the electron tunneling microscope. **Here is the story of how I discovered the fallacies of Modern Physics and the truth that each of us creates zero-point and quantum waves to define our reality and control our bodies.**

Within these pages you are beginning a journey that will help you achieve greater success in life. What you learn will give you greater control over the forces and materials you work with to create your life. This book was written to provide you with a road map to discover what life is and what it is doing. **Reading the book is not important; looking at life is the important thing.** By noticing that you can see things in your own universe when your body's eyes are closed should demonstrate to you that you are a spiritual being capable of creating zero-point energy. The pictures that you see with your mind's "eye" are created with zero-point energy. This book gives you lots of evidence of zero-point energy and the thought universe to confront and learn from. By confronting life you will understand and control it. The goal of this book is to bring each reader to a point where he is better able to understand his own life and thus solve his own problems.

I will attempt to show through pictures and quantum theory that men are spiritual beings composed of nothing, an absolute stillness, who create the illusion of everything in the Physical Universe by means of the production of zero-point energy and quantum wave structures. Through the manipulation of these waves an astonishing improvement in man's health, wealth and joy of living can be achieved. F. Lee Aeilts

Dr. F. Lee Aeilts

INTRODUCTION: UNRAVEL THE MYSTERIES OF INFINITY!

Niels Bohr, the man who mapped the atom, proclaimed that "Anyone who is not shocked by quantum theory has not understood it." Similarly, the theory that man is a spiritual being composed of nothing but capable of creating everything is a shocker to many people. Both theories work; they help us to understand our world and improve it.

JZ Knight speaking at the Orbs Conference in Sedona Az, May 2007. Notice how expanded, transparent and thin the large orb is. It may be Ramtha's creation. The smaller orb may belong to Ramtha's channel, JZ Knight.

TABLE OF CONTENTS

ACKNOWLEDGEMENTS	VI
PROBLEMS WITH MODERN PHYSICS	IX
INTRODUCTION: UNRAVEL THE MYSTERIES OF INFINITY!	X

COSMIC MECHANICS

1) MY FIRST EXPERIENCE 1
2) THE DESERT UFO 8
3) THE HESSDALEN LIGHTS 9
4) SPACE SHUTTLE DISCOVERY UFO (STS-48) 10
5) MUHNOCHWA 14
6) ANCIENT TECHNOLOGY 16
7) EVIDENCE OF CURRENT ALIEN OPPRESSION OF MANKIND 20
8) THE MECHANICS OF 21 GRAM DEMONS 34
9) THE MECHANICS OF DEMON ORB PHOTOS 42
10) COSMIC PULSE OF LIFE 47
11) THE MECHANICS OF FOO FIGHTERS 50
12) THE MECHANICS OF THE PHYSICAL UNIVERSE 52
13) THE MECHANICS OF ENERGY 53
14) THE MECHANICS OF LIGHT 54
15) THE MECHANICS OF MATTER 55
16) FAULTY MECHANICS OF THE BIG BANG THEORY 57
17) THE MECHANICS OF REALITY DISTORTION 59
18) THE MECHANICS OF GRAVITATION 61
19) DO PLANETS ACTUALLY TUG ON PASSING BODIES? 63
20) THE PIONEER ANOMALLY 64
21) MECHANICS OF A HELIUM BALLOON 65
22) THE MECHANICS OF CONSERVATION OF ENERGY 68
23) WAVE MECHANICS 69
24) HOW ARE ZERO-POINT WAVES DIFFERENT FROM PHYSICAL WAVES? 77
25) THE ANATOMY OF A WAVE 84
26) THE SPEED OF A WAVE 89
27) THE DOPPLER EFFECT 91
28) TRAVELING WAVES VS. STANDING WAVES 93
29) BELL'S THEOREM 94
30) CREATION OF THE PHYSICAL UNIVERSE 97
31) DOES QUANTUM MECHANICS APPLY TO HUMANS? 98
32) SOLID UNIVERSES 99
33) GRAVITY 103
34) MECHANICS OF ENERGY FLOWS 106
35) ZERO-POINT WAVES CARRY INFORMATION 107
36) THE MECHANICS OF HYDROGEN 112
37) THE GHOST IN THE MACHINE 115
38) THE MECHANICS OF SPONTANEOUS HUMAN COMBUSTION 117
39) ORBS; SMOKING GUN EVIDENCE OF IMMORTALITY 121

40)	THE MECHANICS OF NEAR DEATH EXPERIENCES	123
41)	WHAT ARE 21 GRAM ORBS?	124
42)	THE MECHANICS OF HYPNOTIC INDUCTION	126
43)	EARTHLIFE IS AN ELABORATELY CONTRIVED ILLUSION	129
44)	THE MECHANICS OF EVIL	130
45)	WHY CAN'T WE SEE ORBS WITH OUR EYES?	141
46)	THE MECHANICS OF OUR PRISON PLANET	142
47)	THE MECHANICS OF COSMIC BEINGS?	146
48)	YOUR MOST VALUABLE POSSESSION IS YOUR UNDERSTANDING	154
49)	THE MECHANICS OF ENERGY PICTURES	155
50)	WATER HOLDS A MEMORY	158
51)	DNA PHANTOM EFFECT	162
52)	SPOOKY ACTION AT A DISTANCE	165
53)	INTENTIONS CHANGE DNA	165
54)	THE MECHANICS OF DNA AND RNA UNRAVELED	166
55)	THE MECHANICS OF THE FIELD	173
56)	OUR MINDS HAVE BEEN COMPROMISED	179
57)	CAN AN ORB THINK, FEEL AND PERCEIVE?	179
58)	CAN AN ORB BE DAMAGED?	182
59)	THE MECHANICS OF AGRIGLYPHS	183
60)	HOW DO COSMIC BEINGS COMMUNICATE?	185
61)	THE MECHANICS OF ORB CREATION	185
62)	HOW COSMIC BEINGS GET BODIES	187
63)	HOW TO PHOTOGRAPH ORBS	188
64)	WHY IS IT BENEFICIAL TO UNDERSTAND ORBS?	189
65)	BIOPHOTONICS – QUANTUM CONTROL OF BODIES	190
66)	DR. FRITZ-ALBERT POPP	192
67)	THE MECHANICS OF PHOTONS	199
68)	THE MECHANICS OF ATOMIC SPECTRA	204
69)	QUANTUM PROPERTIES OF LIGHT	213
70)	EXPOSING THE DREAM WORLD WE BELIEVE TO BE REAL	218
71)	WHAT DOES FAITH IN YOURSELF HAVE TO DO WITH REALITY?	218
72)	ENERGY MEDICINE	222
73)	JOHN OF GOD -- HEALER	229
74)	PRODUCTION OF EFFECTS	231
75)	THE MECHANICS OF SPACE	232
76)	KINESIOLOGY	234
77)	HOW THE QUANTUM VACUUM COMMUNICATES INFORMATION	236
78)	THE GAME WE PLAY IS LET'S PRETEND	238

79) WHAT OUR ORBS DO TO US	239
80) THE SECRET OF THE UNIVERSE	242
81) THE MECHANICS OF SPACE RESONANCE	248
82) THE NEW PARADIGM OF DR. MILO WOLFF	256
83) APPLICATIONS OF SPACE RESONANCES	266
84) EXAMPLES OF SCALAR WAVES	288
85) CONCLUSIONS	288
86) THE END	291

Similar to the merkabah Dr. Aeilts saw in his first experience except for the gold hue; his was white. Notice how the density of the ridges decreases with the distance from the center. This is a ball of condensed energy. Its energy can be calculated by Einstein's famous formula, $E=mc^2$.

1) MY FIRST EXPERIENCE

In 1988 I saw strange lights (above) fly near Groom Lake, Nevada at Area 51. (Satellite photo-below) It is about 90 miles north of Las Vegas. I was researching UFOs with Sean David Morton in Tikaboo Valley, at the  mailbox on Mailbox Road, (see above) watching the sky over Groom. Groom Lake is America's traditional testing ground for black budget aircraft such as the U-2. The vast Groom Lake air base  cannot be seen directly because of the intervening Jumbled Hills, (left, as seen at the far end of Groom Road), but what we recorded that night gave us the data to unravel the secret of the flying lights, which I now refer to as merkabahs. The Hebrew word **merkabah** (הבכרמ "chariot", derived from the consonantal root *r-k-b* with general meaning "to ride") is used in Ezekiel (1:4-26) to refer to the throne-chariot of God. The Bible is replete with references to higher vibrational life forms interacting with mankind. For example, there is Ezekiel's chariot of fire and the pillar of smoke by day and fire by night that lead the Israelites for 40 years through the desert.

We were about to see some of these celestial fireworks as Sean set up his video camera pointed north toward the town of Rachel and started it recording. Shortly, two lights (Like the three orbs shown above and the gold orb – ours was white, not gold) started forming up over Jumbled Hills to the south. Sean saw his big opportunity for a UFO breakthrough so he jumped in his little red sports car and took off down Mailbox Road leaving a cloud of dust in his wake. About half a mile down the road a couple of Area 51 security vehicles cut him off and confronted him at gunpoint in the middle of the dirt road. Sean asserted his right as a US citizen to be on Bureau of Land Management property. But the security guards told Sean to get back in his car and go back. Sean protested until he saw a little red laser sighting beam jumping around on his chest. At that point he "saw the light" and reconsidered, saying something like, "I think I hear my mother calling. Could I get you guys a beer from the Little Ale' Inn?" Sean was always the funniest and gutsiest guy around. But, back to the lights, we were all so transfixed by the motions of the lights that we forgot about the video camera recording the blackness of the night in the opposite direction.

The lights gradually got brighter and brighter until they were each about as bright as a 200 watt electric light bulb. They lit up the hillsides beneath them faintly. One of them slowly rose straight up in the air about a hundred feet, stopped, flew to the right about a hundred feet, stopped, descended a hundred feet, stopped, and then returned to its starting point. It did a perfect silent square loop the way no airplane or helicopter could ever do. Then, the orb that did the square loop started scanning the area. It put out a beam to the west, like a radar beam and started "looking" as it swept counter-clockwise toward the south and then around toward the east. I was beginning to get pretty nervous about this anomaly and I tried to somehow duck when the beam came around to me but there was no place to hide and the beam penetrated me and identified me. Everyone in our group was identified. Don't ask me how I know that. It was just a thought projection and a mapping of the area. It wasn't anything that happened in the Physical Universe, it was just a spiritual event. My best guess is that those beings in the lights checked out the area by directing a zero-point beam to scan the area.

After doing this preliminary flight routine and area scan, the two merkabahs shot off across the desert south toward Las Vegas and disappeared from view. None of us knew the significance of the lights, where they came from, what their purpose was, if they were military or civilian or how they were formed.

When Sean got home he reviewed the video tape just to hear the "oohs!" and "aahs!" of the people who were witnessing the lights. To our surprise, the video showed little firefly-like objects flying by the camera lens while the two lights were growing brighter. That's when it dawned on me that millions of tiny unseen beings were flying in from all directions to join up in two gigantic assemblies that had the combined illumination power of two 200 watt light bulbs. Sean's camera just caught the beings coming in from the north. It was pitch black outside and none of us saw anything in the air. But the camera, having a CCD (left - Capacity Coupled Device) that is sensitive to infra-red wavelengths longer than 400 nanometers picked up the single beings in

flight. I wondered if our eyes had been purposefully genetically engineered to be insensitive to the lower infra-red range so that humans would not be distracted by the life forms that swarm around them and watch. Maybe they do more than just watch. Maybe they influence us somehow. I determined to unravel this puzzle. Only animals can see these beings as they move around us. You may have noticed a dog tracking something with its eyes and barking at it mysteriously when you couldn't see the object of its attention. You may have noticed your TV or VCR turn on or off when no one was touching the remote control. These are all signs of one or more beings with orbs nearby emitting infra-red energy.

On a Sunday evening shortly after that I was watching 'Sightings' and saw a UFO fly across the crater rays of Tycho. It was just a white dot moving at about 7,000 miles per hour across the surface of the moon. Some amateur astronomer had mounted a video camera to his telescope and just happened to catch the flight. In view of the fact that NASA didn't find anything on the Moon during the Apollo program, I wondered where it came from. I don't mean to infer that you should distrust official government sources; I simply mean to insist upon it! You'll notice that the crater rays on the moon don't originate from the center of the craters. Look at them with your binoculars or get a book on astronomy. They come from the mountains on the edge of the craters. They are tangent to the craters and they go out straight to a smooth, perfectly round crater that looks like a hopper, or receiving terminal for minerals mined in the mountains. Mining has been going on for centuries on the Moon and a little of the white ore blows off the sides of the ships each time they make the trip to the processing plant. This operation results in the crater rays we see today. Those rays weren't created by the impact of a meteor in spite of what the

university professors say. You can even see parallel rays coming from Tycho, the crater which has the largest mining operation. They were formed by east and west bound traffic. The ships flying west always took the route on the north side to avoid the incoming, east bound ships which took the southern route. Anyone can see this with binoculars from the Earth, but humans are lied to so consistently that they never catch on. Above is a photo of Apollo 11 astronaut Edwin Aldrin. Notice how well the astronaut is illuminated by the sun from <u>behind</u> him. Notice the strange reflections in Edwin's dark shadow. Where did that light come from? Notice the shadows of the rocks and objects about Edwin and the spotlight shining on the astronaut taking the picture that you can see in Edwin's helmet. The angles and directions of the shadows show that the "sun" was located about 50 feet away inside the studio, not 93,000,000 miles away. This is one of the thousands of photos taken in the NASA studio. Search the worldwide web for NASA photos and you'll find overwhelming evidence of doctored moon pictures.

Don't trust me, go look for yourself.

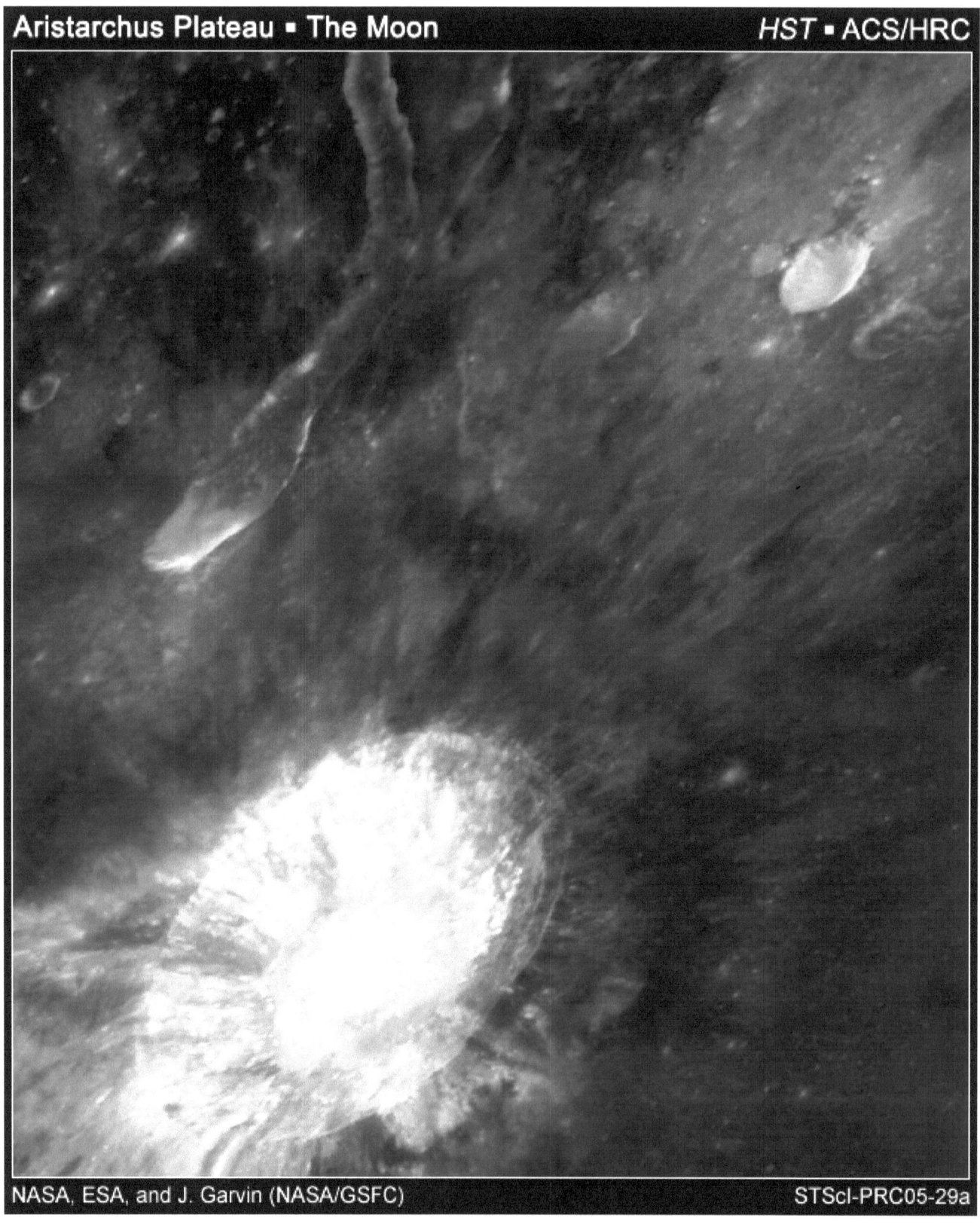

NASA astronauts reported that they found a gray lifeless moon, and they published hundreds of pictures to prove it. They all said that there was no atmosphere on the moon, but take a look at the Hubble Space Telescope image above. Notice the out-gassing from the crater. What do you think? Notice how bright the inside of the crater is compared to the outside. There could be an atmosphere on the moon.

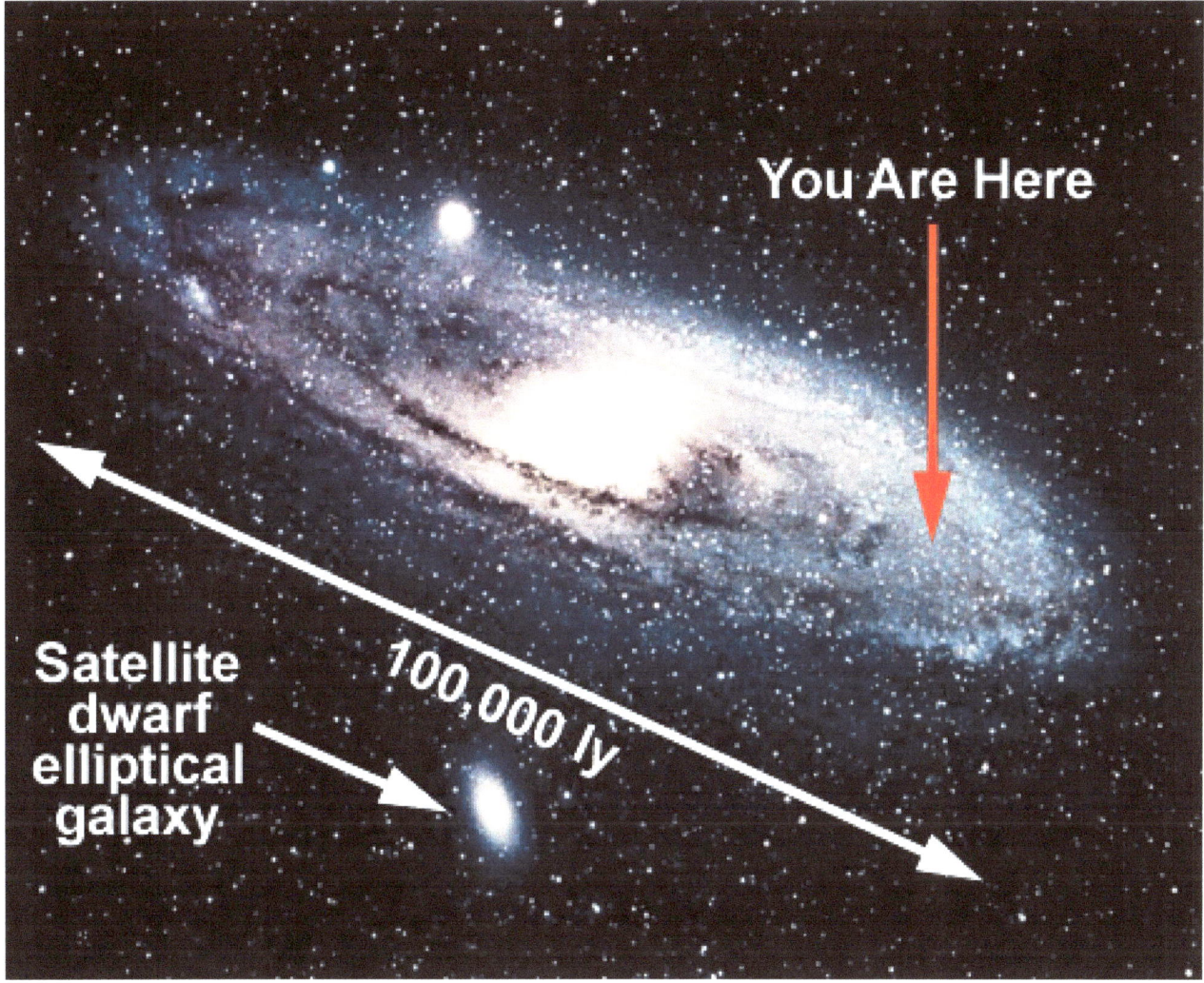

If you'll look closely at NASA's Apollo photos you'll see reflections from lights in the studio ceiling, and the black backdrop for the studio set they used in Houston to create those stunning pictures. There are no stars painted on that backdrop. You'll notice that there is no dust on the lander, nor blast crater formed under it by the rocket engine prior to touchdown. There were some actual photos taken from orbit but all the pictures of men on the surface of the Moon in space suits were produced in a NASA studio. We actually went to the moon, but our men didn't land on it. NASA faked part of the Apollo mission to create a fish story that would fulfill President Kennedy's goal of putting a man on the moon and safely bringing him back. It also secured billions in funding for the space program and gave America a brief technological edge over Russia which launched the first satellite, Sputnik.

Now we have a legitimate space shuttle program and NASA has published on the internet a voluminous list of strange lights and sightings that were seen by astronomers since the days of Galileo. Strings of moving lights have been seen in the crater Plato, and the crater Eudoxus has displayed long lines of light. For thirty years Mount Piton, in the northern section of Mare Imbrium, has been sending out beams of light that look like beacons. Many astronomers and scientists are acknowledging the strange sightings but they keep their professional aloofness by calling them "transient lunar phenomenon reported by the lunatic fringe."

I've given you too much data already that contradicts the world view you've accepted all your life. If you woke up one morning and found that you had been put on a strange island, your first project after getting food and water would be to try to find out where you were. This is what has happened to all of us. We have all been born on an island in space: Spaceship Earth. We are all passengers on a cosmic voyage. Astronomy is the process of finding out where we are -- our position in space and time in the unfolding history of the universe. Everyone can know for themselves. You don't have to ask NASA or the university professors or the Pope. You can look.

It is not the policy of the government for people to know the truth. The truth would reveal the vicious crimes and terrorism committed in the name of government. Joseph Goebels, the Nazis Propaganda Minister, wrote, "The lie can be maintained only for such time as the state can shield the people from the political, economic and/or military consequences of the lie. It thus becomes vitally important for the state to use all of its powers to repress the dissent. For truth is the mortal enemy of the lie and thus by extension, the truth becomes the greatest enemy of the state." Hitler actually staged an attack on his own German Parliament building in February 1933 and blamed it on the Russian Communists. Then he began to legislate away the German people's civil rights in the guise of protecting the German homeland. The German people, as naïve and apathetic and manipulated as Americans are today, supported the Third Reich in their war and genocide.

2) THE DESERT UFO

Another example of this kind of activity which isn't seen by our eyes but which is picked up by the CCD is this **desert UFO**. Jon gave the following story: "This was witnessed while I was home in the Arizona Desert. - slightly northeast of Phoenix/Mesa on Dec 26, 2006. As my family and I were driving down a secluded dirt road in the middle of the desert, my mom leaned out the window to snap a quick photo of the jeep ahead. There was a lot of commotion, bumping of the

jeep, noise etc. Therefore her attention would not have been focused on the sky above. Upon arriving home and unloading the images onto the computer, we noticed that one of them had a hazy white disk-shaped object in the sky. It was almost as if it was flying at a high velocity due to the way it produced a rather odd purple haze around it. I am not a physicist but, was it really atmospheric burn-off? None of the other clouds around it had this coloring.
(Photo: Charlotte Gamel)

Of course, I thought it was awesome that she captured this on "camera". As a result, I immediately took the image, and inverted it to find out if it was real - and not some scratch on the lens or some piece of dust. From my knowledge, there is no way this was not something in the sky.   Things to Note:
1. No pictures before or after have aberrations.
2. My mother would not have altered the photograph. She is not the type to do that.
3. No other clouds surrounding the object have a purple haze.
I have attached the original along with a non-original which I analyzed. My mother has given me permission to re-distribute this image. Thanks. --Jon G. "

3) THE HESSDALEN LIGHTS

In an excellent article on her Earthfiles website, Linda Moulton Howe reported on the utter bafflement of Norwegian scientists over mysterious lights appearing in the valley of **Hessdalen, Norway**. They look very similar to the Merkabahs I have seen. The investigators noted that sometimes the lights would dim out and disappear to the naked eye, but they could still be seen in an infra-red night vision scope. That indicates that they can lower their frequency below the visible electromagnetic spectrum.

Linda writes of the research there: "The results can be broken down into two groups: 95% are thermal plasmas and 5% are unidentified solid objects. The plasmas emit long wave radio frequencies and strangely, their temperatures do not vary with change in size or brightness." She quotes the scientists' research summary:"1) most of the luminous phenomenon is a thermal plasma; 2) the light-balls are not single objects but are constituted of many small components which are vibrating around a common barycenter (center of mass); 3) the light-balls are able to eject smaller light-balls; 4) the light-balls change shape all the time; 5) the luminosity increase of the light balls is due to the increase of the radiating area. **But the cause, and the physical mechanism with which radiation is emitted, is currently unknown**."

Next came this report of UFOs in space which was similar to other reports I had seen.

4) SPACE SHUTTLE DISCOVERY UFO (STS-48)

On September 15, 1991, between 20:30 and 20:45 Greenwich Mean Time, the TV camera located at the back of Space Shuttle Discovery's cargo bay was trained on the Earth's horizon while the astronauts were occupied with other tasks.

A glowing object suddenly appeared just below the horizon and "slowly" moved from right to left and slightly upward in the picture. Several other glowing objects had been visible before this, and had been moving in various directions. Then a flash of light occurred at the lower left of the screen; and the main object, along with the others, changed direction and accelerated away sharply, as if in response to the flash. Shortly thereafter a streak of light moved through the region vacated by the main object, and then another streak moved through the right of the screen, where two of the other objects had been. Roughly 65 seconds after the main flash, the TV camera rotated down, showing a fuzzy picture of the side of the cargo bay. It then refocused, turned toward the front of the cargo bay, and stopped broadcasting.

Clip from NBC broadcast on UFOs

COSMIC MECHANICS

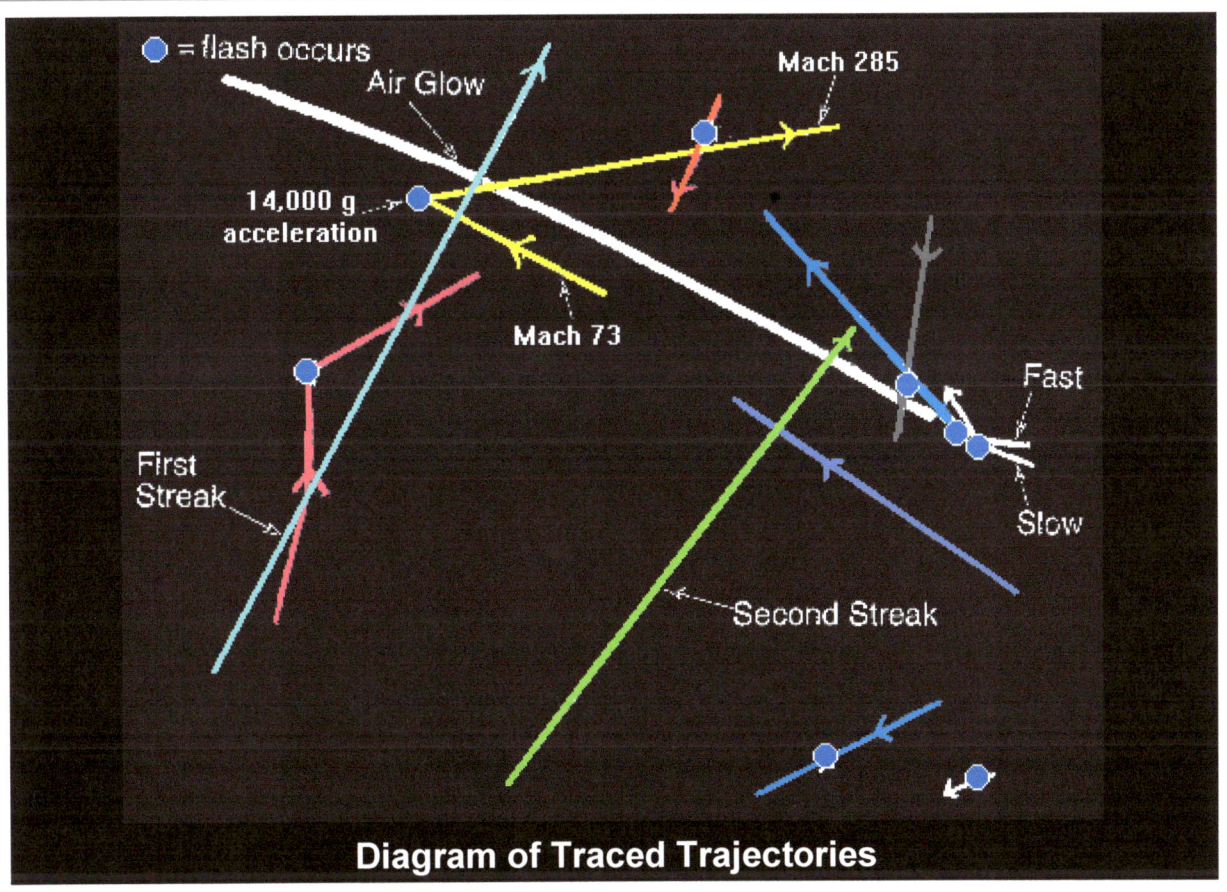

Diagram of Traced Trajectories

- The distance from the Discovery to the Earth's horizon is 2,757 kilometers.
- The UFO's speed before accelerating into space is 87,000 kph (Mach 73).

- **Three seconds after the light flash, the UFO changes direction sharply and accelerates off into space at 340,000 kph (Mach 285) within 2.2 seconds.**
- **Such an acceleration would produce 14,000 g of force.**

One of the most spectacular video footage of a UFO encounter was taken by cameras on board the Discovery space shuttle, STS-48, on 15 September, 1991. The video sequence was picked up live by a number of amateurs who were directly monitoring the transmissions. The material has been shown in news broadcasts worldwide.

The video shows several small bright objects maneuvering on screen, apparently interacting with one another in a complex fashion. Skeptics have usually insisted these are merely shots of some of the many small ice particles which inevitably end up in orbit with every space shuttle.
UFO investigators were quick to dispute this interpretation, and US scientist Richard C. Hoagland soon conclusively demonstrated the objects were actually large in size and many hundreds of kilometers away from the shuttle.

One UFO in particular appears to rise up from below the Earth's dawn horizon and can be clearly seen emerging from behind the atmosphere and the 'airglow' layers. It is certainly in orbit around the Earth, some distance out in space, and traveling quickly.

A sudden, bright flash of light is then seen to the left of the picture, below the shuttle. The UFO then turns at a sharp angle and heads out into space at very high speed. Two thin beams of light (or possibly condensation trails) move rapidly up from the Earth's surface towards where the UFO would have been if it had continued in its original orbit.

Careful analysis of the video shows: SEE DIAGRAM ABOVE

The light flash and light beams (or contrails) that shoot into space have variously been described as a ground-based attempt to disrupt or destroy the UFO. Hoagland interprets the incident captured by the Discovery's video camera more specifically as a "Star Wars" weapons test against a Star Wars drone (the UFO). Other UFO investigators prefer to describe it as a Star Wars attempt against an extraterrestrial UFO. Whichever version you prefer, the technology implied is most certainly impressive - at least of Star Wars caliber.

More recently, from New Zealand, investigators have reviewed the video and corrected the actual time it was taken. They have found that the UFO incident was recorded over Australia and not the Philippine islands as was originally thought. Discovery's trajectory had already taken it across Surabaya in Java and above the Simpson Desert, Western Australia.

The UFO is first picked up coming over the horizon when the shuttle is close to Lake Carnegie, WA. Later, the light flash and one contrail can be tracked back to Exmouth Bay near the North West Cape military facility. A second contrail can be tracked back to the Pine Gap military facility in central Australia.

US investigators have been asking their Australian counterparts to provide further information which they don't have and which they probably can't get. All the information we have on this incident so far comes from the US or New Zealand.

And, of course, there are Australia's stringent secrecy laws to contend with. The scenario was probably captured on video purely by chance. Along with other UFO incidents recorded on video by NASA, this material has contributed significantly towards NASA's recent decision to discontinue live television transmissions from space.

Reprinted from the September 1995 issue of UFO REPORTER, the quarterly publication of UFO Research (NSW), P.O. Box Q95, Queen Victoria Building, Sydney, NSW 2000, Australia). The video footage itself (including a detailed investigation of the incident) can be found on the tape "Hoagland's Mars Vol 2 : The UN Briefing (Extended Version)."

* Diagram & text based on/borrowed from: http://www.mufor.org/sts48.htm

Next, I saw this "atmospheric plasma" photo and this headline in an Indian newspaper that seemed to offer clues to the technology of the mysterious orbs. The article comes from the Times of India August 20, 2002, which illustrates the conflict between electronic beings (with orbs) and animal body civilizations:

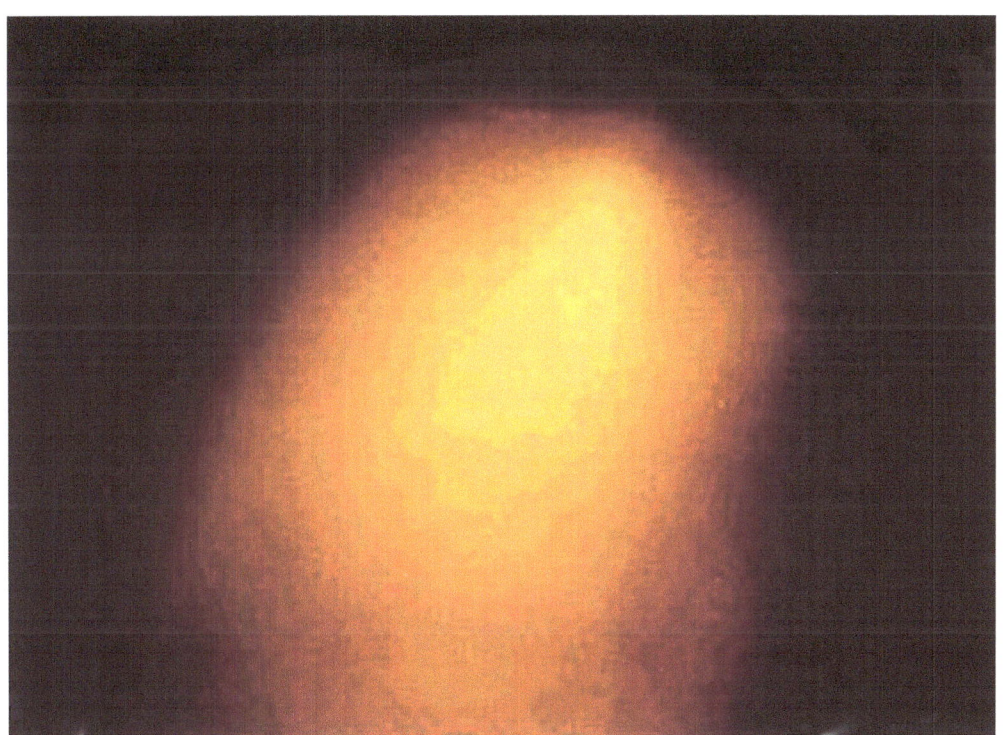

5) MUHNOCHWA

INDIA CALLS IN X-FILES AGENTS TO UNMASK FACE-SCRATCHING ALIEN

by Catherine Philp
UFO sightings have sparked hysteria and riots, says our reporter.

A mysterious flying object said to attack sleeping villagers has sparked mass hysteria and rioting across the north Indian state of Uttar Pradesh. Police shot dead one man and injured 12 others when a mob of hundreds stormed the police station in Barabanki, demanding protection against what they believe is an alien assailant terrorizing villages. The object, described as a flying sphere emitting red and blue light, is said to strike in the middle of the night, leaving victims with burns or scratches on their faces and limbs, and earning it the name the muhnochwa (face-scratcher).

At least seven unexplained deaths in the area have been attributed to the muhnochwa, sparking panic among villagers who blame police for not providing enough protection. Officials have suggested a raft of explanations, from an alien invasion to a new and unknown breed of insect.

Perhaps the most bizarre theory was that of Police Deputy Inspector General K. N. D. Dwivedi, who said that the assailant was a genetically engineered insect introduced by "anti-national elements" from outside India to cause mayhem.

In common Indian parlance, this is taken to mean that it was the work of the Pakistani spy agency, the universal scapegoat for all unexplained Indian woes. That theory has not won over many believers.

Villagers across the region no longer sleep outside, as they usually do during the sweltering summer heat and long power failures, fearing that they will be easy prey for the muhnochwa. In some villages the entire population is squeezing into the headman's house for the night, seeking shelter and safety in numbers. Having lost faith in the police, villagers have formed nocturnal protection squads.

In Shanwa village, where the attacks are said to have started, men patrol all night, banging drums and shouting slogans to frighten off intruders, such as: "Everyone be alert. Attackers beware."

Residents have dismantled television aerials and taken satellite dishes down from their roofs, fearing that they may attract the mysterious object. Even radios have fallen silent at night under self-imposed blackout.

The Times of India reported that the national intelligence bureau was sufficiently concerned to send its own agents, like Mulder and Scully from television's X-Files, to investigate the "alien" invasion.

After listening to villagers' descriptions of the muhnochwa, the agents constructed their own replica from the base of a mixer-grinder, fitted with colored lights, and hoisted it onto a pole in an attempt to entice the extraterrestrial. Then they waited.

At 1.05 am they were rewarded with a flash of light "like a photocopier", which repeated three times. A videotape was said to show a flash of light passing across the screen. The agents concluded that the villagers were right and that they were indeed experiencing an extra-terrestrial invasion.

Local doctors, however, have dismissed the phenomenon as mass hysteria, saying that most of the injuries have been self-inflicted by panicked villagers, evoking memories of the "monkey man" hysteria in Delhi last year.

At least three people died jumping from roofs and dozens more were injured during the mystery simian's (resembling an ape or monkey) two-week reign of terror before officials dismissed it as a mass delusion and sightings petered out.

6) ANCIENT TECHNOLOGY

I looked into ancient technologies to get an idea of what was going on around me. There seemed to be a class of beings that were smarter and faster and more able than humans. I wanted to find out about them so I checked into the technologies of ancient Sumer and Egypt to see what clues I could gather about the mysterious orbs. Ancient Egyptians had a better grasp of cosmic mechanics than exists on Earth today. Egyptians got their technology from Sumer in Mesopotamia, now known as Iraq. Sumer is also the source of our sexagesimal mathematics based on the number sixty. We have a 12 month calendar, twenty-four hour day, sixty minute hour, sixty second minute, 360 degree circle and 12 inch foot.

The Sumerians have been reported to be reptilian life forms with six fingers and six toes that set up a biosphere between the Tigris and Euphrates rivers in what is now known as Iraq. The Reptilians did genetic research to crossbreed their bodies with the indigenous bodies of early mankind to produce a working class of slaves – us. The Bible calls this place the Garden of Eden where Eve was beguiled by a serpent to partake of the forbidden fruit with Adam and was cast out of the biosphere. Eve wanted the forbidden fruit whenever and with whomever she choose, not just as directed by the genetic scientists. Adam decided to go along with her plan to escape the biosphere and gain their independence. Even today all human embryos go through a reptilian stage in the fourth week of their development and once in a while a baby is born with six fingers and six toes. Below is a boy awaiting surgery to remove the extra digit and make him "normal." Notice how beautiful he is in his present condition.

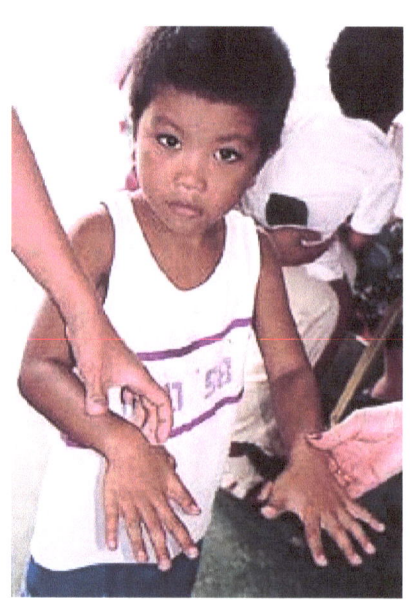

People of the Waorani tribe of Ecuador have six fingers and six toes. During the 1970s, one Dr. James Yost had made substantial efforts to learn about the mysterious Waorani people and their unknown origins. In the process, Yost found some facts about the Waorani that were disturbing:

COSMIC MECHANICS

Unravelling the secrets of Waorani culture, Jim made some amazing discoveries. He found that the Waorani had inherited more than just the physical attributes of their reptilian ancestors. Due to the influence of their reptilian reactive mind, or orb, they maintained the highest levels of homicide ever recorded in the annals of human history. Fully fifty per cent of all deaths in the preceding five generations had been the result of homicide as the Waorani engaged in a continuous and deadly internal vendetta, pursued mostly at night, in spearing raids. No death, it seemed, whatever the cause, went unavenged. Furthermore, the Waorani were even reputed to kill by spearing any, although only a few instances have been proven, of their old people who no longer had the means to support themselves; and they practiced infanticide, either strangling unwanted or malformed babies with vines, or burying them alive.

Waorani males with six fingers on their hands and six toes on their feet. Images from WAORANI: The Saga of Ecuador's Secret People: A Historical Perspective

The Waorani, though extremely violent, also exhibited superior genetics, being some of the healthiest people in the world: "Medically, the Waorani turned out to be something of an enigma: they had no trace of cancer; no cardiovascular disease; no high blood pressure; no allergies; and none of the known diseases familiar to us." Not only were they physically superior in many ways, some even had six fingers on each hand and six toes on each foot — characteristics, as we have seen, of the giants. It would appear that the giants of old that had invaded Ecuador had integrated with the native Asiatic population, creating a hybrid race of people that were in many ways physically superior — and extremely violent.

The Bible gives us more detail about the six fingered people. (2 Samuel 21:20)
*There was war at Gath again, where there was a man of great stature **who had six fingers on each hand and six toes on each foot, twenty-four in number**; and he also had been born to the giant (**Heb.** Raphah or **Rapha**, product of the **Rephaim** living within the Canaanite Philistines).*
The Bible confirms this record by repeating it:
*(**1 Chronicles 20:6**)*
*Again there was war at Gath, where **there was a man of great stature who had twenty-four fingers and toes, six fingers on each hand and six toes on each foot**; and he also was descended from the giants [Heb. **Rapha**].*

Zecharia Sitchin translated ancient Sumerian clay seals which describe a race of beings who came from space to establish the most technologically advanced civilization in history. You can read about it in *The 12th Planet (The Earth Chronicles, Book 1)*. Their technology was brought to Egypt to build the pyramids. You can read about Egypt's fascinating technology in *The Giza Power Plant: Technologies of Ancient Egypt* by Christopher P. Dunn. You can still see the remains of one of their power generation systems today in the great pyramid of Cheops on the Giza plateau near Cairo. Bedouins of the first century found metal pipes in the north and south walls of the great pyramid. Their records say that whereas the Romans use lead to make their pipes of circular cross section, the Egyptians used copper to make pipes with a rectangular cross section.

Electrical Engineers know that copper pipes of rectangular cross section are wave guides, and these pipes had the proper dimensions to transmit hydrogen microwave frequencies. The copper waveguides and the power generation device have been pulled out of the Kings Chamber. Some clever Germans, led by Rudolph Gantenbrink, discovered the purpose of the pyramids by back-engineering the remaining artifacts. National Geographic sent a robot (above) built by iRobot of Boston with a camera up the tunnels leading from the Queen's Chamber and found a flotation valve and chemistry equipment for generating hydrogen gas. The light gas rose up into the King's Chamber to be used by a MASER. That's <u>M</u>icrowave <u>A</u>mplification through the <u>S</u>timulated <u>E</u>mission of <u>R</u>adiation. The so-called "sarcophagus", carved out of one solid piece of granite, housed the electronics. One end is rounded, not for decoration, but to focus the microwave energy into a feed horn conducting the energy up the wave guide in the south wall where it was beamed by parabolic high-gain antennas to the desired location.

COSMIC MECHANICS

These are the 'flying machine' hieroglyphs from the Temple of Seti I in Abydos. They are evidence of wonderful ancient Egyptian technological ability. You can see a helicopter built like a CH-54 Skycrane and a huge airship lifting something resembling a building stone for the pyramid. On the right there seems to be an illustration of a microwave tower more sophisticated than the one shown here to beam energy to the airships.

What powered the maser? Researchers found several huge monoliths balanced securely on granite anchors above the maser in the Kings Chamber. These anchor blocks connect with the base of the pyramid which has been set in the bedrock beneath the sand. If you examine these monoliths near the ceiling of the Kings Chamber you will notice that they support nothing. They are highly polished granite beams on the bottom and sides, but oddly they have a rough, unfinished texture on their top surfaces. Their purpose is revealed when you tap each beam with a rubber mallet and hear their low pitched vibration – each one tuned to the same frequency! The Egyptians really understood wave mechanics! They chipped away at the top surfaces of the beams to tune them to the resonant frequency of the earth so that long wavelength vibrations from the Earth's core would couple to the beams, causing them to vibrate sympathetically and generate electric power. The electricity came from electrodes placed on the beams to pick up the piezo-electric charge that the granite crystals produced from the vibrational bending. The electrical energy thus produced was used to power the laser to produce the microwave beam which shot into space. A parabolic reflector, similar to today's radar antennas, could direct the energy for domestic use anywhere in sight of the pyramid. That includes the Moon when it was up. I think that space shuttles departing from the space port at Eden caught the pencil tight beam over Egypt and used its energy to launch them into space, thus obviating the need to carry fuel for the trip into space. Once out of Earth's gravity well, a short burst of energy could propel the craft across light years of space. This is such a smart solution to space travel compared to our rocket monsters which we destroy on each launch. It is also millions of times more efficient, resourceful, safe, and conservative than our space shuttle system.

Egyptologists claim that the pyramids were the tombs of the pharaohs but that is totally false information. Not one mummy was ever found in any of the pyramids nor any inscriptions of praise or eulogy or artifacts like you find in the tombs of the pharaohs in the Valley of the Kings. They also say that the word "pyramid" means sex, from *pyra* – fire, and *mid*, in the middle. But it actually describes an energy generation station, having a literal fire in the middle, the high powered MASER.

Egyptologists have long been puzzled why the master architects who built the pyramids of Giza didn't place the three in a straight line. They are all meticulously oriented to true North, and are built to engineering standards that are inspirational, yet they seem to be placed in an odd pattern. Wouldn't it be more aesthetic to place them in line? No, since the pyramids were used for an

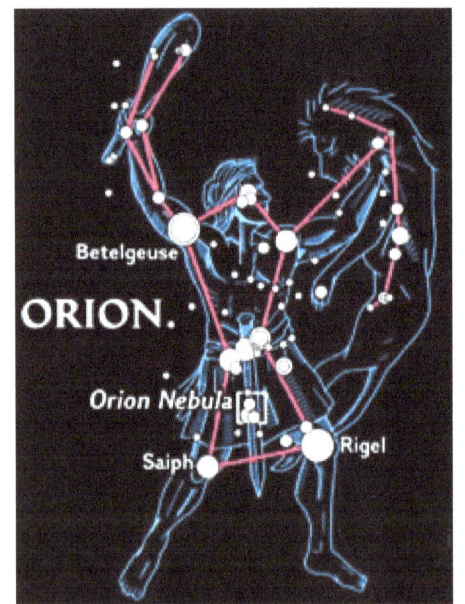

initial approach fix to the spaceport at Eden between the Tigris and Euphrates rivers, they were offset from a straight line in the same proportion as the belt stars of the Constellation of Orion. From space, one could see the Constellation of Orion on the Earth's surface at night because temples (still in place) were built at locations in Egypt that corresponded with the other stars in Orion, and the Nile represented the Milky Way. When the fires of the altars of the temples were ablaze at night space travelers could easily see the Constellation of Orion in Egypt and find their way to the spaceport.

7) EVIDENCE OF CURRENT ALIEN OPPRESSION OF MANKIND

Do you believe there is some kind of conspiracy in the world preventing you from being all you can be? Have you noticed that the American educational system is the richest in the world yet falls behind third world countries in results? Does anyone not want us to have the best technology and education in the world? For instance, is Psychiatry dumbing our kids? It is prescribing Ritalin and other psycho-active drugs to younger school children. It has a history of brutal, inhumane treatments such as electro-shock, drugs and lobotomy that injure, maim and destroy people in the guise of help. Psychiatry is not a science and has no proven methods to justify the billions of dollars of government funds that are poured into it and psychiatric theories that man is a mere animal have been used to rationalize the wholesale slaughter of human beings in World Wars I and II.

Is Big Pharma helping or hurting America? None of the 45,000 people who have sued Merck over heart attacks or strokes suffered after taking Vioxx have received payments from the company.

Is an alien cartel in control of the White House? Why can't Congress stop the Iraqi war that President Bush conned the American people into now that the lies have been uncovered? Is there something we're not supposed to know?

In 1947 when an alien disk was reported to have crashed in the New Mexico desert near Roswell, the American military covered it up claiming it was a crashed weather balloon and the bodies recovered were claimed to be dummies used in Army tests. After years of plausible denial and subterfuge the secret of the cover-up has finally been revealed.

Lt. Walter G. Haut was the Roswell Army Base Public Information Officer in 1947. He denied the cover-up his whole life but when he died in December 2005 he left a notarized affidavit confessing the part he played in the cover-up and the details of the spacecraft and alien bodies he had seen. A copy of the affidavit first appeared in the June 2007 book *Witness to Roswell* by Tom Carey and Don Schmitt.

THE DAY AFTER ROSWELL, by Col. Philip J. Corso   A Former Pentagon Official Reveals the U.S. Government's Shocking UFO Cover-up. This is a landmark expose' firmly grounded in fact. This book ends the decades-old controversy surrounding the mysterious crash of an alien disk near Roswell New Mexico in 1947. Backed by documents newly declassified by the Freedom Of

Dr. F. Lee Aeilts

Information Act, Col. Philip J. Corso, (Ret.), a member of President Eisenhower's National Security Counsel and former head of the Foreign Technology Desk at the US Army's Research and Development Department, came forward to reveal his personal stewardship of alien artifacts from the Roswell crash. He spear-headed the Army's reverse-engineering project that led to the development of today's integrated circuit chips, fiber optics, lasers, super-tenacity fibers and "seeded" the Roswell alien technology to giants of American industry.

Did the Army hide the evidence of the crashed disk in the interest of National Defense? Or did some select few obtain for their personal aggrandizement the secrets of alien technology? Or was it somehow not in the "custodians" best interest to expose humanity to the presence of alien beings on the planet? What is the agenda of the "custodians" regarding these secrets? Let's look at the evidence.

Evidence of oppressive secret "custodians" operating on earth:
1. The World Trade Towers weren't brought down by jet aircraft which flew into them on 9/11, but by demolition charges put into position in advance. It was an inside job. The 9/11 Commission covered all this up.

2. President George W. Bush used the public emotion and anger over the 9/11 incident to incite war in Iraq after drumming up phony evidence of Sadam Hussein developing weapons of mass destruction.
3. When the truth came out about how President Bush conned America, the Republican hawks were voted out of office and the anti-war Democrats took over control of Congress. But that didn't change the course of the war that the "shadow government" was pursuing. The

astonishing revelation was that the Democrats still couldn't stop the war or even cut funding for it. The shadow government who was making money by waging war was more powerful than the populace who wanted to end it. Also, when the White House pushed legislation making the previous CIA illegal monitoring of American's telephone calls and email legal, the Democrats were powerless to stop it. It makes one believe that democracy is two wolves and a lamb voting on what to have for dinner. Where does this secret power

come from?

It will help if you know that during WWII the Union Banking Corporation of New York City was a Nazi money laundering bank. It helped finance Hitler's rise to power. It was exposed for harboring millions of dollars of Nazi money. It was seized for violations of the Trading With The Enemy Act. The Director and Vice President of the Union Bank was Prescott Bush, father of President George H. W. Bush and Grandfather of President George W. Bush. Now you understand better

  the moral and political position of the Bush family which stands at the head of American government today.

The idea that a small, wealthy, ruling elite – the "reptilian custodians" – controls America seems to be well supported by the facts. A disproportionate amount of America's resources is controlled by a handful of its 300 million population. According to a 1983 study by the Federal Reserve Board, a

mere 2 percent of US families controlled 54% of the nation's wealth. The majority – 55 percent – had zero or negative net worth. And today the difference between the haves and have-nots is growing. In 1998 the average worker's median pay – adjusted for inflation – was one full dollar below the 1973 hourly rate, and the situation is worsening.

The United States stands alone as the world's preeminent power. So who really controls the United States and hence, the world?

Everyone's heard how "they" own the vast majority of resources, manipulate stocks, control prices and avoid paying taxes. "They" also maintain monopolies over energy, medicine, armaments, and manufacturing by suppressing new technologies. And "they" wield undue influence over the news media and world governments with their control of multinational corporations as well as private organizations such as England's Royal Institute of International Affairs, the Council on Foreign Relations and the Trilateral Commission. "They" also belong to secret societies such as the Illuminati, Skull and Bones, Knights of Malta, the Bilderbergers and the inner circles of Freemasonry. What secrets do they possess that allow them to assume the role of the custodians of planet Earth? And behind those few powerful men is there a master race, such as the ancient serpent kings, directing the show? If so, what is their agenda and why is it so important that the vast majority of humans remain out of the loop; in ignorance of the forces and motives of the custodians? What are the mechanics of oppression? Is this truly a prison planet?

It is clear to the most casual researcher that secret societies not only do exist but have played key roles over the centuries. The Rothschilds had the crown heads of Europe in debt to them and this included the Black Nobility dynasty, the Hapsburgs, who ruled the Holy Roman Empire for 600 years. The Rothschilds also took control of the Bank of England. If there was a war, the Rothschilds were behind the scenes, creating the conflict and funding both sides. The international bankers are not moral beings – a characteristic of the ancient serpent kings.

When Wellington's revitalized army defeated Napoleon at Waterloo in June, 1815, news of the victory was rushed to England by Rothschild couriers bearing their well known and untouchable red pouches. The Rothschild messenger arrived a full day ahead of Wellington's own courier. Knowing of his capacity for early intelligence, all eyes on the London Stock Exchange turned to Nathan Rothschild, who, appearing despondent, ordered the sale of his stocks. Certain that Wellington had been defeated, a selling frenzy began, with the end result being that Nathan Rothschild's agents were able to buy up a hefty majority of Britain's debts for only a small portion of their true value. Much later, Nathan Rothschild commented on his act by saying, "It was the best business I have ever done." This is typical of the thinking of a reptilian – sneaky, crafty, deceitful, greedy, criminal and loathing. There are many beings like him on the planet today. They are anti-social personalities.

The imprint of the secret societies can be found in every war and conflict of the 20th century. The historic record is unmistakable. The same society members turn up in every instance – passing from father to son, business partner to close associate, fraternity brother to brother. The same old secret society faces keep returning to power. The mass media seems unconcerned and the public is asked to believe that this is all shear coincidence – simply a case of the most competent man for the job. But nothing happens in politics by chance. If something happens, you'd better believe it was planned that way.

The world's deepest secrets all lead back to Sumer in Mesopotamia, now known as Iraq, whose ancient records state that the nation was founded by astronauts who came from the stars in great ships. This high tech culture appeared about six thousand years ago, left its impact, and then vanished without a trace. It was reported to be a six fingered, six toed, reptilian race with astonishing scientific skill and our sexagesimal mathematics based on the number 60 came from them. Bits and pieces of their prehistoric knowledge survived in various esoteric forms through secret societies such as the Mystery Schools of Egypt and the schools of Pythagoras. These little-understood groups passed along not only religious concepts such as reincarnation but also real knowledge in mathematics, architectural design, construction, astronomy, agronomy and history. Sumer was the apex of scientific achievement and subsequent Egyptian, Greek, Roman, English and American cultures have not been able to duplicate it.

Dr. F. Lee Aeilts

What is the greatest secret that the Illuminati, the Knights Templar, the Freemasons, the Bilderbergers, Skull and Bones, MJ-12, the Catholic Church or the reptilians have to hide?

1. Is it that over the years more than half of the priests in the Roman Catholic Church have been pedophiles and sexually molested millions more children than have been discovered?
2. Is it that disembodied spirits operating through paranormal channels are taking control of the planet and establishing the New World Order?
3. Is it that there is a UFO cover-up and the US government is cooperating with space aliens to take over the planet?
4. Is it that the Roman Catholic Church never did have divine authority to rule the Earth? Rather than participating in ecumenical studies to determine which traditions have the more factual basis, the Church instead attempted to eradicate any challenge to its authority by the most violent and murderous means, such as the Inquisition.
5. Is it that the reptiles use public institutions to suppress scientific advancement like they did in 1631 with the conviction of Galileo Galilei for heresy?
6. Is it that secret societies see war as the basic social system which has governed most human societies of record and still does today?
7. Is it that the reptilian gods of antiquity became tired of toiling on the earth to obtain gold and precious minerals for their intergalactic corporation so they got the bright idea of inter-breeding with the indigenous humanoids to produce a dumb class of workers to labor in their place? According to ancient Sumerian clay seals translated by Zecharia Sitchin an initial effort to retrieve gold from the Persian Gulf by a water extraction system proved inadequate for their needs. So gold mining operations were set up in Africa and South America. Proof of such early gold mining has come from scientific studies conducted for the Anglo-American Corporation, a leading South African mining corporation in the 1970s. Company scientists discovered evidence of ancient mining operations which were dated as far back as 100,000 B.C. Similar ancient mine excavations have been found in Central and South America. One Sumerian text

reported "When the gods [Anunnaki], like men, bore the work and suffered the toil – the toil of the gods was great, the work was heavy, the distress was much." So with genetic manipulation they created a new race in their own image as reported in Genesis of the Bible. The genes given to man were inhibited in a conscious effort to prevent any competition from the new human race. According to Genesis 3:5, the very first order of the *Elohim* was that man--in the allegorical form of Adam and Eve—was to remain ignorant lest "ye shall be as gods." (King James) Man was also given a short life span. The Epic of Gilgamesh stated, "Only the gods live forever under the sun, as for mankind, numbered are their days, whatever they achieve is but wind." As bluntly stated in the Bible, Adam and his progeny were not destined for a life of ease, but one of hard work and survival at the hands of their "Lords." "The term that is commonly translated as 'worship' was in fact *avod* – 'work,' " stated Sitchin. Ancient and biblical man did not worship god, he feared and worked for him. The Sumerian texts make it clear that the Anunnaki, the Serpent Kings, treated their created slaves poorly, much like we treat domestic animals we are merely exploiting – like cattle. Slavery in human societies was common from the first known civilizations. The Anunnaki were vain, petty, cruel, incestuous, and hateful, worked their slaves hard and had little compassion for the plight of humans. The Anunnaki set up their work force and disappeared from Earth using mind control, heavy taxation, international banking and secret societies to divert the tax surplus to off-world projects. That's why the serpents had to leave; that's why Moses couldn't see the face of God on Mount Horeb, only his back parts. That's why ancient Egyptian depictions of the gods were given heads of birds or animals instead of serpents. That's why sightings of UFOs are debunked by the shadow government. If the serpents are nowhere to be seen, they wouldn't be suspected of being part of this scheme. When the half breeds turned out to be smarter than was expected, they had to be dumbed down with television disinformation, drugs, psychiatry, religion, inefficient educational institutions, corrupted language systems, false sciences, destructive technologies and corrupt political systems. With the advent of the machine age and a plentiful supply of oil for energy, mankind had more discretionary time and affluence. To prevent him from using his new discretionary time to discover the source of his suppression, he was kept busy fighting artificially created wars, litigating with one another in a flawed judicial system, killing, stealing and conning each other with a flawed criminal justice system. The only system that was perfected and worked flawlessly through all of mankind's violence and woes was the international banking system which expanded gloriously with every war and conflict. The reptilian "custodians" were winning at man's expense!

8. That last one sounded pretty explosive! What could be more shocking than that? What is the planet buster that has to be kept under wraps at all costs? Here it is for your enjoyment:

The greatest, most diabolical secret on the planet, that any self respecting serpent would die for rather than divulge is the fact that

each human is a divine spirit operating a physical body and is capable of creating matter, energy, space and time out of nothing. They are gods in their own right!

We will never die and our integrity is more important than our present body. We use to be Gods of our own universes, but now we are fallen and no one will help rehabilitate us. We have total amnesia when we should have total recall. No one is encouraging us to throw off our chains. We've only been given inhibitions. No one has allowed us to know our true potentialities. We've been oppressed and mind controlled by alien beings for the purpose of making us unknowing slaves. We are entertained and defrauded by television each day while the alien agenda relentlessly unfolds. When word gets out that we've been mind controlled, oppressed, held down and used by a reptilian race to further their own nefarious purposes, mankind will break out of his chains. It will be endgame for the reptilians. That's why the reptiles know that this secret must never see the light of day on planet Earth. That's also why every human must discover this secret and make it known to every other human. Together we can undo the mental blocks to our individual memories and use existing technologies to get out from under the oppression of the reptilians. To learn more, read *Rule by Secrecy* by Jim Marrs. (ISBN 0-06-093184-1)

THE LOSS OF CIVIL RIGHTS AND THE RISE OF SLAVERY

Modernly, the reptilians, whose life span exceeds 3,000 years, have been busy behind the scenes directing a few men to bring the nations of earth to their knees and set up a one world totalitarian government to accomplish the reptilian agenda. I know this sounds science fiction and unreal. You'd never hear about this on television, the reptilian controlled media, which would rather have you entertained and ignorant rather than intelligent and aware. To get more information see *Zeitgeist* at www.zeitgeistmovie.com.

President John F. Kennedy said, "The very word "secrecy" is repugnant in a free and open society and we are as a people inherently and historically opposed to secret societies, to secret oaths, and to secret proceedings. We are opposed around the world by a monolithic and ruthless conspiracy that relies primarily on covert means for expanding its sphere of influence, on infiltration instead if invasion, on subversion instead of elections, on intimidation instead of free choice. It is a system that combines military, diplomatic, economic, scientific and political operations. Its preparations are concealed, not published. Its mistakes are buried, not headlined. Its dissenters are silenced, not praised. No expenditure is questioned, no secret is revealed." President Kennedy tried to get the US government to print its own money and return the US Constitution to power but he was assassinated to avoid a confrontation with the international bankers.

In 1775 The American Revolutionary war began. The primary issue behind the revolt was the fact that King George III of England outlawed the interest free independent currency the colonists were printing and using for themselves, forcing them to borrow money from the Bank of England at interest, immediately putting the colonists in debt. Founding Father, Benjamin Franklin said, "The refusal of King George III to allow the colonies to operate an honest money system, which freed the ordinary man from the clutches of the money manipulators was probably the prime cause of the revolution." The reptilians already had control of Europe. It was this oppression that the early pilgrims and colonists were trying to get away from when they came to America.

In 1783 America won its independence from England. However, its battle against the central bank and the reptilian custodians associated with it had just begun. Thomas Jefferson (1743-1826) said, "I believe that banking institutions are more dangerous than armies. If the American people ever allow private banks to control the issue of currency…the banks and corporations that will grow up around them will deprive the people of their property until their children wake up homeless on the continent their fathers conquered."

In 1907, J. P. Morgan exploited his influence to spread rumors about the insolvency of competitor banks and established a run on the banks which resulted in a great depression. Fredrick Allen of Life Magazine wrote, "The Morgan interests took advantage…to precipitate the panic of 1907, guiding it shrewdly as it progressed." However, a congressional investigation headed by Senator Nelson Aldrich recommended a central bank so the panic of 1907 would never happen again. Thus the Federal Reserve Act was put through, unconstitutionally putting the money power into the hands of a private bank.

In 1913, Woodrow Wilson became president and discovered the shadow government. He stated, "Our great industrial nation is controlled by its system of credit. Our system of credit is privately concentrated. The growth of the nation, therefore, and all of our activities are in the hands of a few men…who necessarily, by very reason of their own limitations, chill and check and destroy genuine economic freedom. We have come to be one of the most ruled, one of the most completely controlled and dominated governments in the civilized world – not a government by free opinion, no longer a government by conviction and the vote of the majority, but a government by the opinion and duress of small groups of dominant men."

After the passage of the Federal Reserve Act, Congressman Louis McFadden proclaimed, "A world banking system has been set up here…a super state controlled by international bankers…acting together to enslave the world for their own pleasure. The Fed has usurped the government." The public was told that the Federal Reserve was an economic stabilizer and that inflation and economic crises were a thing of the past. As history has shown, nothing could be further from the truth.

On May 7, 1915 the SS Lusitania was sent into the war zone carrying US passengers and war munitions. It was torpedoed by a German submarine killing 1700 people. It was a set-up by international bankers to draw America into WWI. The Imperial German Embassy had enough heart to advertise in the New York Times warning Americans not to travel by ship in the war zone. During the war 323,000 American lives were lost and J. D. Rockefeller made $200,000.00. The war cost Americans $30 billion dollars which was borrowed at interest from the Federal Reserve Bank.

In 1920, the Fed called in huge numbers of loans causing runs on local banks, bankruptcy, and a collapse of the financial system. Congressman Charles Lindberg stated, "Under the Federal Reserve Act, panics are scientifically created. The present panic is the first scientifically created one, worked out as we figure a mathematical equation." Mayer Amschel Rothschild, Founder of the Rothschild banking dynasty, said "Give me control of a nations money supply and I care not who makes its laws."

In 1913 the bankers pushed through congress the Federal Income tax which was completely unconstitutional and the 11th Amendment which made it possible was not ratified by a majority of the states. It is a testament to how dumbed down, unaware and naïve the American people were and still are. A tax of 35% of the average workers income is confiscated each year by the bankers who print money and lend it to the government at interest as if it were valuable property. No law exists which requires the American people to pay federal income tax, yet they lie down, keep their mouth shut and take the oppression. Our Federal Income Tax payments go directly to the privately owned bankers who use it to offset our huge national debt which they have created. This calamity occurred because Americans as a group and individually did not take responsibility for their economic life. So the bankers stepped in and took advantage. Similarly, American's do not take responsibility for their reactive minds (orbs) so the reptilians step in and take advantage of them with mind control operations.

In 1929, International bankers called in their loans to stock brokers causing another 10,000 banks to fail. The International bankers purchased their receivables at pennies on the dollar and eliminated banking competition. It was the greatest robbery in American history. Then the Fed continued to contract the money supply causing one of the greatest depressions in history. The Fed bankers purchased huge corporations for pennies on the dollar. Congressman McFadden stated, "It was a carefully contrived occurrence. International bankers sought to bring about a condition of despair, so that they might emerge the rulers of us all." McFadden tried to impeach the board members of the Federal Reserve but he was killed by poisoning before he could carry it out.

In 1937 the international bankers passed legislation to seize the gold of US citizens under penalty of fines and imprisonment. Americans went along with this because they were told it was needed in the war effort. After that the gold standard was abolished, making paper currency unstable and subject to inflation. The Federal Income Tax is nothing less than the enslavement of the entire country. The Federal Reserve banking system gives the bankers the ability to cause war and profit thereby. The most lucrative event for the international bankers is war because it forces the nations to borrow money from them at interest. They care not what carnage they create or how many lives are lost in the struggle. This is the alien mindset as dictated by their orbs. Getting products to their corporation is their mission and it doesn't matter what cost to human life and happiness. They are not moral beings. They treat us as we would treat our cattle.

On Dec 7, 1941 the Japanese attacked Pearl Harbor which brought America into WWII. The attack was provoked by F. D. Roosevelt and his shadow government. Roosevelt deliberately got the American Navy out of the way of the Japanese task force so he could proclaim his "Day of Infamy" and start borrowing from the Fed. Roosevelt came from a long line of New York bankers. His uncle Fredrick was on the original Federal Reserve Board.

In 1964 the Gulf of Tonkin incident was reported to be an attack by a Vietnamese PT boat on a US destroyer. It was a lie promoted to get America into the war in Vietnam. It was a completely staged event. Later the Secretary of Defense Robert McNamara said, "It was a mistake." The war in Vietnam was never meant to be won, just maintained. The White House issued executive orders to never chase North Vietnamese troops over their border and not to bomb strategic sites without prior approval from the Joint Chiefs of Staff. The war caused 58,000 American and 3,000,000 Vietnamese deaths. The Fed made a terrific profit.

On Sept. 11, 2001, a couple of planes flew into the World Trade Towers followed by demolition explosives which brought the buildings down in seconds. It was a staged pretext for war that got Americans angry enough to follow President George W. Bush into the War on Terror based on false intelligence data. The White house claimed that Saddam Hussein was working with the Taliban and building weapons of mass destruction in Iraq. This was a bald-faced lie drummed up to get Americans into a war with Iraq which could not be won. During the war Americans lost their civil liberties with the passage of the

Patriot Act and the Military Tribunals Act. This gave power to the reptilian dominated government to search American homes when the occupants are not there without a warrant, arrest Americans with no charges revealed and be detained indefinitely without the aid of a lawyer if Americans are "suspected" of being a terrorist. These new laws are being put into place to prevent Americans from defending themselves from what is coming on the alien agenda. History repeats itself – in Feb. 1933, Adolph Hitler staged an attack on his own German Parliament Building and blamed it on the Russian Communists. Thus he started WWII by misleading the German people and manipulating them. Hitler was involved in secret spiritualist groups which instructed him in oppressive foreign policy and genocide. This is the temperament of the reptilians. It isn't what you'd expect from a human. There are people guiding your life unbeknownst to you. Television is a big part of it. Drugs, psychiatry, the military/industrial complex and our educational system with its scientific materialism are also involved. The last thing the reptiles want is a conscious, informed public. They don't want reports of crashed discs, photographs of paranormal orbs or thousands of dead cows in the Midwest with body parts surgically removed. Also, it just makes their job tougher when crop circle phenomena in England are published and alien anti-gravity drive equipment becomes available for humans to handle. I think that there are a few hidden reptiles on earth to direct the secret societies and handle the liaison for space ships that come and go to carry materials on and off the planet. I believe most of the international bankers have reptilian-like orbs because they came from reptilian cultures but they are incarnated in human bodies, so it is a great shock to us when they create wars that are so unconscionable to us. They look like us, humans. But they talk like George W. Bush who is dedicated to defraud America into self destruction, war and bankruptcy. He is also making the military/industrial complex and international bankers rich while robbing the working class of the fruits of their labors. I think that there are a few hidden reptilian life-forms on earth also because of reports I have heard. I once saw a video image of a reptilian person on television who had been caught on tape by a British surveillance camera in the middle of the night. Later the tape was debunked as showing a man in a Halloween costume. But I saw what I saw.

It is clear now that the reptiles intend to bring all humans into subjection by making them subject to oppressive laws and implanting them with micro-chips. If any human should get out of line the chip that carries the monetary, medical, vital signs and personal identification data will be turned off. This will result in a human population utterly malleable and indefensible in the hands of the heartless reptiles. This is our future if we don't wake up our fellow humans soon. We will be reborn over and over again on planet earth only to be slaughtered forever in our relentless search for freedom.

It may appear that I have diverged from the theme of this book, *Cosmic Mechanics*, by delving into dirty politics. But you'll never understand the mechanics of the cosmos if you don't get your wits around the interplay of spiritual beings with their diabolical orbs.

8) THE MECHANICS OF 21 GRAM DEMONS

Those of a religious bent believe in life everlasting for the faithful, a continuation of the life force that reaches far beyond the limitations of mortal flesh. In such belief systems, death is not an end but a transformation: though people shed their corporeal selves at the moment of demise, that which made them unique beings lives on into eternity. This is the cosmic being or spirit; an entity described in the dictionary as "The immaterial essence, animating principle, or actuating cause of an individual life." Yet as much as we believe in the concept of "spirit," this life spark remains strictly an article of faith for most people. As central as it is to our perception of ourselves, it can't be seen or heard or smelled or touched or tasted, a state of affairs that leaves some of us uneasy. Without the spirit, dead is dead. But if it could be proved to exist, a great deal of anxiety over what happens to us when we die would be resolved.

This is exactly what Dr. Duncan MacDougall of Haverhill, Massachusetts was attempting to do in 1907. The doctor erroneously thought that the spirit was material and therefore had mass, ergo a measurable drop in the weight of the deceased would be noted at the moment this essence parted ways with the physical remains. The belief that human beings are possessed of spirits, or ghosts, which depart their bodies after death and that these spirits have detectable physical presences were around well before the 20th century, but claims that spirits have measurable mass which falls within a specific range of weights can be traced to Dr. MacDougall's experiments.

Unfortunately, Dr. MacDougall had no idea of what he was measuring so he jumped to some confusions. He thought that he was measuring the human soul, or animating spirit, (which has no form or weight) when actually he was measuring the departure of the 21 gram orb.

Dr. MacDougall, seeking to determine "if the psychic functions continue to exist as a separate individuality or personality after the death of brain and body," constructed a special bed in his office "arranged on a light framework built upon very delicately balanced platform beam scales" sensitive to two-tenths of an ounce. He installed upon this bed a succession of six patients in the end stages of terminal illnesses (four from tuberculosis, one from diabetes, and one from unspecified causes); observed them before, during, and after the process of death; and measured any corresponding changes in weight. He then attempted to eliminate as many physiological explanations for the observed results as he could conceive: The patient's comfort was looked after in every way, although he was practically moribund when placed upon the bed. He lost weight slowly at the rate of one ounce per hour due to evaporation of moisture in respiration and evaporation of sweat.

"During all three hours and forty minutes I kept the beam end slightly above balance near the upper limiting bar in order to make the test more decisive if it should come."

"At the end of three hours and forty minutes he expired and suddenly coincident with death the beam end dropped with an audible stroke hitting against the lower limiting bar and remaining there with no rebound. The loss was ascertained to be three-fourths of an ounce. "

"This loss of weight could not be due to evaporation of respiratory moisture and sweat, because that had already been determined to go on, in his case, at the rate of one sixtieth of an ounce per minute, whereas this loss was sudden and large, three-fourths of an ounce in a few seconds. The bowels did not move; if they had moved the weight would still have remained upon the bed except for a slow loss by the evaporation of moisture depending, of course, upon the fluidity of the feces. The bladder evacuated one or two drams of urine. This remained upon the bed and could only have influenced the weight by slow gradual evaporation and therefore in no way could account for the sudden loss.

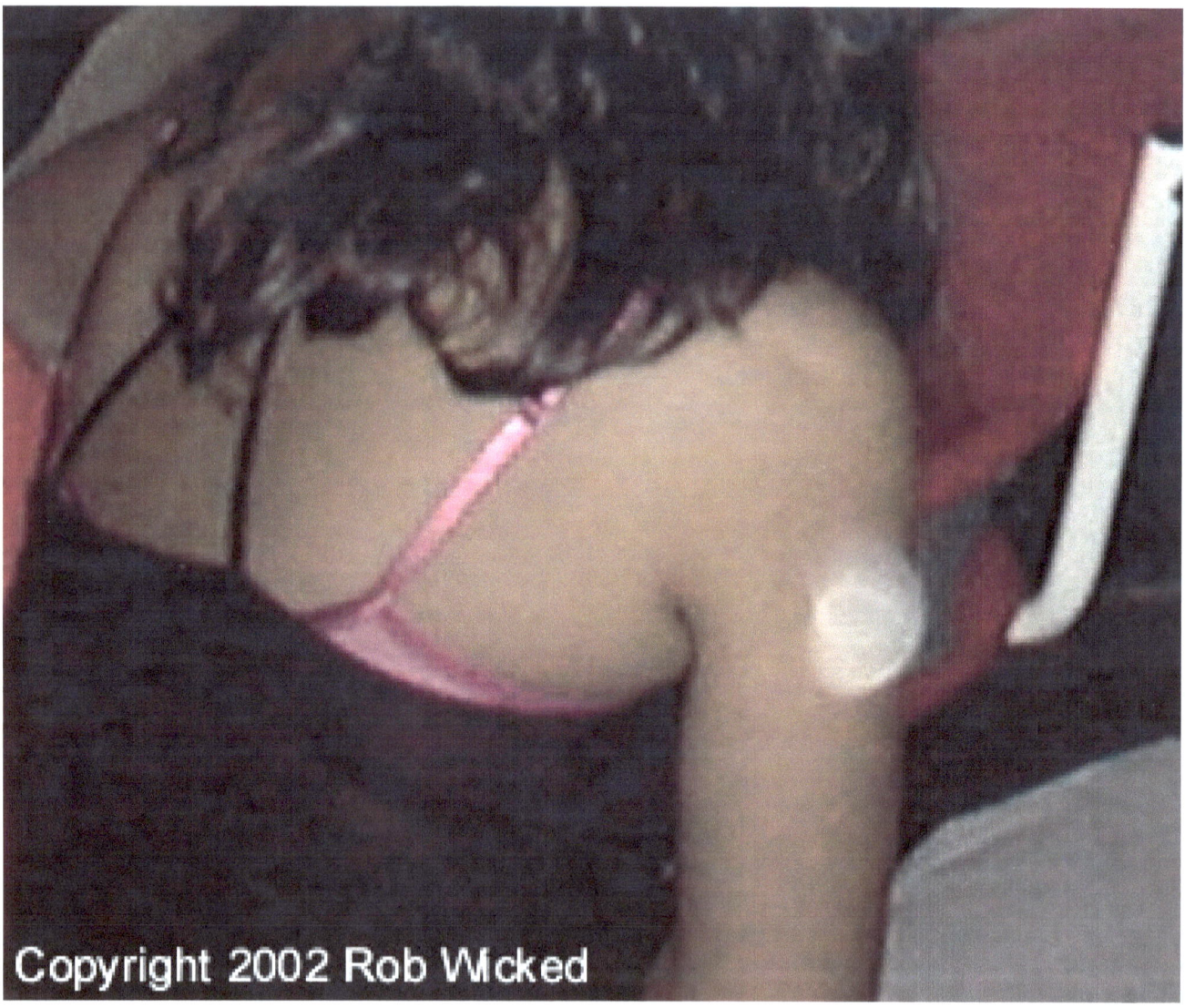

There remained but one more channel of loss to explore, the expiration of all but the residual air in the lungs. Getting upon the bed myself, my colleague put the beam at actual balance. Inspiration and expiration of air as forcibly as possible by me had no effect upon the beam. My colleague got upon the bed and I placed the beam at balance. Forcible inspiration and expiration of air on his part had no effect. In this case we certainly have an inexplicable loss of weight of three-fourths of an ounce. Is it the soul substance? How other shall we explain it? "

MacDougall repeated his experiment with fifteen dogs and observed that "the results were uniformly negative, no loss of weight at death." This result seemingly corroborated MacDougall's thought that he had proven his hypothesis that the loss in weight recorded as humans expired was due to the soul's departure from the body, since (according to his religious doctrine) animals have no souls. Actually, animals have no demon orbs. Have you ever seen an animal get revenge? They protect themselves and their kind but there is no eye for an eye, tooth for a tooth, and "he hit me first!" The impulse to hit back when we have been slighted comes directly from our demon orbs. (MacDougall's explanation that "the tests on dogs would be obtained in those dying from some disease that rendered them much exhausted and incapable of struggle" but "it was not my fortune to get dogs dying from such sickness" led author Mary Roach in *STIFF* to observe that "barring a local outbreak of distemper, one is forced to conjecture that the good doctor calmly poisoned fifteen healthy canines for his little exercise in biological theology.")

In March 1907 accounts of MacDougall's experiments were published in *The New York Times* and the medical journal *American Medicine*, prompting what Mary Roach described as an "acrid debate" in the latter's letters column.

Dr. MacDougall admitted in his journal article that his experiments would have to be repeated many times with similar results before any conclusions could be drawn from them: "If it is definitely proved that there is in the human being a loss of substance at death not accounted for by known channels of loss, and that such loss of substance does not occur in the dog as my experiments would seem to show, then we have here a physiological difference between the human and the canine at least and probably between the human and all other forms of animal life. I am aware that a large number of experiments would require to be made before the matter can be proved beyond any possibility of error, but if further and sufficient experimentation proves that there is a loss of substance occurring at death and not accounted for by known channels of loss, the establishment of such a truth cannot fail to be of the utmost importance." Since Dr. MacDougall's death in 1920, many such experiments have been performed on dying humans and animals using sophisticated digital scales. Dr. MacDougall's results have been validated. Dogs do not lose 21 grams at expiration because they are not animated by cosmic beings who have been busy creating demon orbs for 200 trillion years.

I had picked up valuable clues to understanding the demon orbs from Dr. MacDougall's experiments. Above is a photo of ectoplasm building up over a Ouija board. That would be a clue.

This is where the old man sat before he died. What can we learn from this? Is this the evidence of morphogenic fields that remain long after the life force has departed? These strange mists are the product of colliding zero-point energy flows created by a spiritual being.

The demon orbs were somehow related to dead people, as shown above. They are composed of such fine particles that they do not experience wind resistance. I continued to hunt for answers.

This is a photo of a demon orb above the street at Dealey Plaza where President John Fitzgerald Kennedy was assassinated. I'm not proposing that JFK's orb was photographed after all these years at the exact spot where he was assassinated. I don't know. Someone should go to Dallas and take another picture to see if it is still there. There are many reports that the orbs of American Civil War casualties can be photographed to this day on battlefields where they died. When men die violent deaths they sometimes become so dazed that they hang around for hundreds of years not knowing where they are. Countless pictures of paranormal entities truly test both our intellectual limits as well as our eyes to define what it is that we are actually seeing. One picture after another continues to challenge the best of us. They raise our understanding of not only entities, but the implications of our very being as humans.

This photo shows the inside of a church located in Kellnersville, WI, at a funeral of a well respected person, as well as a friend of many, seeking a true path. His name was Casey Holt,

Paranormal Investigator for the Minneapolis/St. Paul area. He was lost at the age of 39, on Dec. 24th, 2002, in an accidental drowning while ice skating out to the middle of a pond to record temperatures of the water at different depths, in Maribelle, WI. It is a bright white orb, floating toward the ceiling on the upper left hand side of the pic, (Casey perhaps?). The picture was taken with a Sony Cybershot digital camera, and is unmodified, from camera to PC. There was no sign of the orb before the flash went off. I would like to thank Casey's family for their permission to use these photos to demonstrate that, "If Ye Seek, Ye Shall Find", and that, "No Mystery Is Closed To An Open Mind." Fellow Truth Seeker, -Roger Colby.

Notice that the condensed ball of mass is composed of the ridges that surrounded the body at the time of death. The violent death caused the being to collapse his space into the little ball. The energy of the ball can be computed with $E=mc^2$. The energy in one 21 gram orb is equivalent to hundreds of thousands of gallons of gasoline.

Two boys, playing around, were caught in this snapshot as they fell to the floor. A demon orb is seen breeching the skull of the boy near the top. Usually the standing waves fill the space surrounding the body but the force of the impact caused them to contract into a little ball in the head.

My father was a male nurse before he became a chiropractor. He told me of an experience he had of carrying an old man who was in his care to a waiting ambulance. He told me the man died in his arms and he could feel the spirit leave. The man's body got a little lighter at that moment. Maybe the orb breaching the skull above was the object that left the old man's body. This was a valuable clue to understanding those mysterious demon orbs. Maybe they assume the natural shape of the heavenly spheres because they have the gravity of mass and distribute their mass according to the Minimum Amplitude Principle the same as planets, moons and stars do.

Three orbs watch a neighbor clear a road of debris from Nathan and Tegan Huff's mobile home near Continental, Ohio. A tornado destroyed the home on Nov. 10, 2002

9) THE MECHANICS OF DEMON ORB PHOTOS

I started hunting for more photos of demon orbs and spiritual phenomena. The photo below shows a being that took off straight up when the flash went off. It was extremely fast. It seems to me that orbs are very defensive. They've been burned, fried and zapped so many times that now they run to hide when the photoflash just begins to ignite.

I next went to Benson, AZ to get some photos of the mystery lights seen there. I didn't get any photos, but while I was in the desert an orb made three passes in front of my girlfriend's face. She thought that she wouldn't have to go chasing after those orbs. They could come to her; and so they did. She said that the orb came in from the left, lit up like a firefly and then flew away. This

COSMIC MECHANICS

A cosmic being can do the "fire-fly" trick. In 1/60 second this being took off when it saw the flash. Notice the segmented trail which is evidence that it is pulsating on and off. 50 segments in 1/60th sec is roughly 3,000 pulses per second. Notice how fast cosmic beings move compared to humans.

happened two more times from the right side. They know what people are thinking and can show themselves in the dark if they have enough energy to glow. Since then I've learned that lots of the beings have lost their power after dwelling for trillions of years in this universe; so they are unable to glow. It isn't that this universe is so hostile but some of the beings that live here, like the reptilian aliens, are incredibly oppressive.

Beings without bodies move infinitely faster than meat bodies. In the Physical Universe nothing travels faster than the speed of light, but in the thought universe nothing travels slower than the speed of light. If an object in the thought universe were to slow down to sub-light speed it would pop into the Physical Universe.

Here are demon orbs over people who are exerting effort.

COSMIC MECHANICS

A woman goes out to a field in the Blue Ridge Mountains of Virginia to attract orbs. She does pretty well until a man decides to help her. Then they scatter. (Circa 2002)

Photos courtesy Larry Jack Schmidt.   Below, an unembodied demon orb is checking out the neighborhood.

Others were getting pictures of orbs so I decided to try my luck. I went to the hospital across the street from my home to see if orbs were busy leaving old bodies and getting new ones. There is a 21 gram orb in the photo (LEFT) next to the yard light of the Kaiser Permanente Hospital in Hollywood. The sidewalk leading to the Emergency Room entrance is on the lower left and the fire truck, with red and yellow lights, that brought in the body is parked along the curb on the right. The being had just experienced a painful death and was very agitated and charged up as it watched the firemen bring its body into the hospital. (Photo by Dr. F. Lee Aeilts)

I was learning that at the moment of death, a being rises out of its host body. It takes its orb with it. It knows immediately that it has been set free and is going home (It is time to report back). These advanced beings need no one to greet them. However, most beings are met by guides and loved ones. This is demonstrated by this next photo (above) I took of the emergency entrance to the Kaiser Permanente Hospital in Hollywood, CA. It shows eleven beings thought to be attending departing spirits. Note that some of them are thin, gossamer, transparent and expansive. Those are the orbs of spirits who have adjusted well and are free of the pain and misery of the circumstances causing their death. The orbs that are opaque, dense, white and small have had their anchor points caved in. In other words, they are in shock, overwhelmed, stressed out by their recent death and it may take many years to adjust to their new condition. They have pulled in the ridges that used to exist around their body and condensed them into a tight little ball of energy.

I've seen other people's videos of orbs making crop circles and cavorting in the sky near Area-51. I've heard very interesting reports from truck drivers who have seen the Marfa lights in Texas and the Tombstone lights in Arizona.

10) COSMIC PULSE OF LIFE

In *The Cosmic Pulse of Life* (1976), Trevor James Constable published a number of photos showing huge airborne amoeboid-shaped creatures which he dubbed "critters". In all cases, he could not see these creatures with the naked eye, but rather caught the images on infrared film or he used an 18A filter over the lens of his 35mm camera to block out all *visible* light and only allow light from the (unseen) extreme ends of what we call the visible light spectrum.

The photo seen above was photographed by Trevor on August 25, 1957 from the Mojave Desert of southern California. Objects which emit heat will stand out on infrared film as lighter shades while colder objects will appear darker. Very cold areas will appear black. This shot is only one of many living atmospheric bioforms captured on IR film during the period from approximately 1957-62 by Trevor and his friend, Dr. Jim Woods, on the early morning desert plateau prior to sunrise.

This photo below was taken by Constable's daughter near a "cloud buster" orgone energy transmitter. It is an excellent example of the photographed "bioform" -- or "sky fish" as Trevor

calls them. **Orgone energy** is a term coined by physician and psychoanalyst Wilhelm Reich for the "universal life energy" that he described in published experiments in the late 1930s.

COSMIC MECHANICS

Trevor usually used a Leica G 35 mm camera for stills and super 8 color film for motion pictures during the late 1950's and 1960's. His 1975 blockbuster book, *The Cosmic Pulse of Life*, included this shot along with other photos of UFO's and pulsating, plasmoidal bioforms caught on IR film or in some instances, on regular high speed color film using an 18A infrared filter.

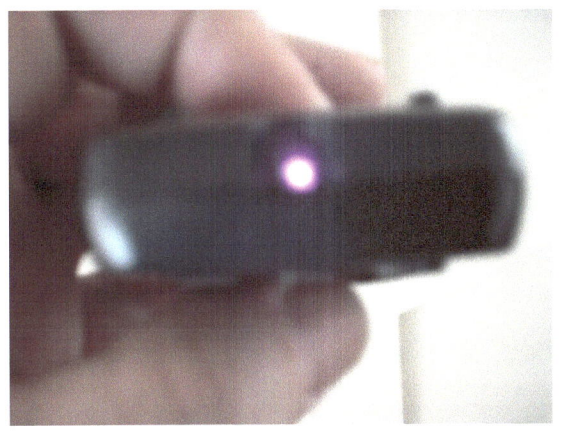

You can test your eyes for infrared sensitivity by looking at the end of your TV remote control as you push the channel select button. You probably won't see any glowing light coming out of the remote. But now turn on your camcorder and press the channel select button on your remote control as you watch the little red flashes on your view screen. The CCD in your camera could "see" it but you could not. It is known that the CCD or CMOS sensors in digital cameras are more sensitive to infrared than human eyes. CCD or CMOS sensors are the elements in digital cameras that are equivalent to film in traditional cameras. Above, the emitter of a remote control is photographed while pushing a button on it. Eyes couldn't see it but the camera did. This picture proves it. It also shows that digital cameras register infrared lights, which the human eye cannot see, as white light.

This UFO photo (above) was taken in March 2005 - A dark field with no lights - Trevor could not 'see' the UFO but it did appear in the photo. According to Trevor Constable: "The so-called "spacecrafts" are just as "aetheric" as the bioforms; as they "are built of matter of the aetheric plane...", and powered by the life-force, what William Reich calls "orgone energy". Their builders are, Constable tells us, a race of intelligent and benign beings, the "aetherians", whose world is the same as our world, except it is of a higher vibratory level. The craft become visible and tangible when they materialize into our own world."

Now I was getting the idea that I was working with an arcane knowledge that most people didn't understand at all. I wondered how I would change if I found out what was going on? I had a burning desire to discover the truth.

11) THE MECHANICS OF FOO FIGHTERS

The next thing I checked out was the Foo Fighters of WWII. **The story went like this:** December 22, 1944: The pilot of the Allied plane was nervous. He was at 10,000 feet, over enemy territory. Somewhere hidden in the black sky there was sure to be German fighter aircraft. He scanned the darkness looking for trouble. Suddenly he saw two large, orange glowing balls approaching him. His radio operator saw them too. They didn't look like enemy fighters, but neither did they look like anything he'd ever seen.

The balls suddenly leveled off and started following the plane. The pilot decided to try and lose them with evasive maneuvers. He put his plane into a steep dive. The objects immediately followed. Next he tried a sharply banked turn. The objects stayed with him. For several more minutes the pilot used his best tricks to lose his pursuers and failed. When he was about to give up suddenly the objects were gone, disappearing suddenly into the night. During the whole incident not a shot was fired.

The above is a typical example of an encounter with a "foo fighter." Toward the end of World War II pilots began reporting seeing strange glowing balls flying around their aircraft at night. The objects seemed to maneuver with great speed and the Allies began to worry that the German's had developed a new weapon with startling capabilities.

The objects were dubbed "foo fighters." because of a popular comic strip at the time, Smoky Stover. Smoky was fond of saying "Where there's foo there's fire" and the objects seem to be fiery, rounded shapes.

Another encounter was described by Major William D. Leet: "My B-17 crew and I were kept company by a 'foo-fighter,' a small disc, all the way from Klagenfurt Austria, to the Adriatic Sea. This occurred on a 'lone wolf' mission at night, as I recall, in December 1944..." Major Leet goes on to note that the intelligence officer that debriefed him and his crew "stated that it was a new German fighter, but could not explain why it did not fire at us, or if it was reporting our heading, altitude and airspeed, why we did not receive anti-aircraft fire."

More encounters with the foo fighters were reported, but none of the objects ever seemed to take any aggressive action, so the idea that they were an advanced enemy weapon was dropped. After the war was over it was learned that German pilots had been seeing the same things and German military authorities had feared an Allied secret weapon.

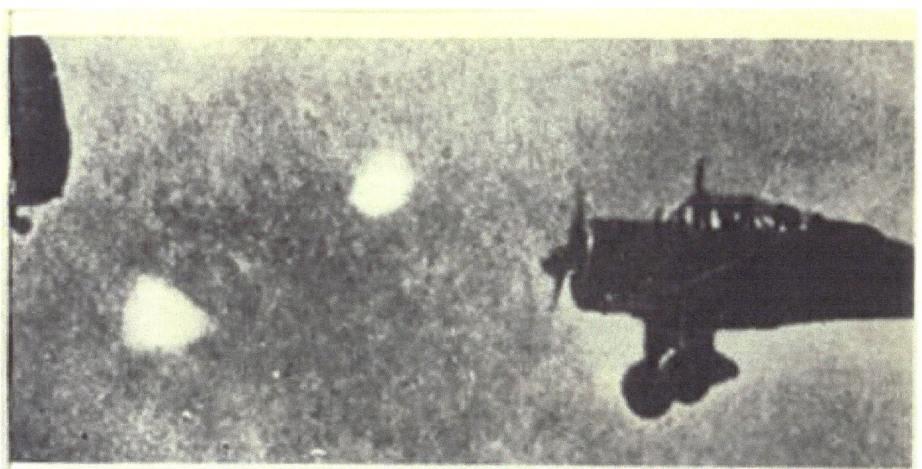

1945. Photo alleges to show two "foo fighters" accompanying a flight of Japanese Takikawa-Kawasaki 98 fighters over the Suzuki Mountains, central Japan. *From Turris & Fusco, "Obiettivo sugli UFO".*

So what were the foo fighters? The military decided they might be an unusual electrical or optical effect (maybe related to ball lightning). They also considered the possibility that the whole thing was in the imaginations of the plane crews who were justifiably nervous under the pressure of flying dangerous war-time missions. No conclusive explanation has ever been found until now. Now it was starting to dawn on me that cosmic beings create orbs as automated mental machines to accomplish their purposes. Also, cosmic beings can unite or join up to build a huge juggernaut that other beings couldn't resist, a merkabah.

I have presented many awesome viewpoints in this book. You are a big being and should be able to have them or not, as you chose. You are actually as big as the number of viewpoints that you can have until ultimately, you are like God, who experiences all viewpoints.

The following is the story of how I figured out what the orbs are doing and how this universe operates. **I also discovered the greatest suppressed secret of all time: that in the distant past you were practically a god and now no one is permitting you to be rehabilitated. Your own abilities are not nearly as limited as you have been made to believe!** For some, this story is the mother of all nightmares. For others it is the hope of eternity.

ALL SERIOUSNESS ASIDE: The shadow government has classified this book SUPER TOP SECRET, MAJESTIC, EYES ONLY, CRYTOGRAPHIC, DESTROY BEFORE LOOKING MATERIAL. For national defense and security reasons, this material must not fall into the hands of 21st century humans. It would terrorize them and mob violence would destroy our cities. Ha!

12) THE MECHANICS OF THE PHYSICAL UNIVERSE

In order to ferret out the truth of the material universe we have to understand its laws. To discover the cosmic mechanics which are at work in the universe it will be necessary to confront intimately the world around us. We really have to look! We all entered this universe and live out our lives within its laws and principles. We agreed to live by the physical laws of the cosmos, from the inescapable law of gravity extending across the universe to the fundamental principles behind the tiniest atoms. These physical laws are orderly and harmonious in spite of the chaos which we see in the heavens after trillions of years of evolutionary development.

As intelligent beings, it is only natural for us to wonder about the world around us, and as spiritual beings it seems reasonable that we should be able to understand it all. In fact, to many it may seem as if we have already arrived at this understanding, with only a few minor details remaining. Isaac Newton gave us an understanding of gravity as an attracting force in nature, and from there many others have contributed to our understanding of light, electricity, magnetism, atomic structure, etc. Our university professors have "proved" that man is an animal which has evolved from the mud they think has conceived it. They don't believe in past lives, yet past lives resulted in their degraded condition. They don't understand their own minds. They think their brain is their mind. They don't have a mental health technology that will clear a person's reactive mind or get rid of their orb. If they can't get rid of 100% of the orbs with their technology, then there is something they don't know about their fellows. If they can't get rid of any orbs, but just make them bigger and meaner, they're a psychiatrist and their stock in trade is drugs, insulin shock, electro-shock and prefrontal lobotomy.

Scientific materialism has removed the superstitious and religious thinking of ages gone by. Now we have the "real" truth which can ferret out the mysteries of the human genome and send space probes out of our solar system. We've come a long way from Stonehenge and the horse and buggy days. The scientific method has finally brought us to a point where we have theories that cover every known observation. But are they correct? We will never understand the technology of UFOs and orbs until we understand our own spiritual nature. Our golden age of understanding has made it possible to invent radio, television, computers, spacecraft, microwave ovens, radar, sonar and deep submergence vehicles. But **men don't understand themselves and their spiritual relationship to the Universe.** Gaining that understanding is what I was after. I would have to look carefully at the world to uncover its secrets. I started to study the physical laws of the universe. First I looked at energy.

13) THE MECHANICS OF ENERGY

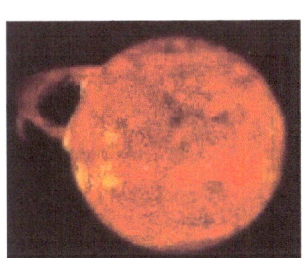

How much do we really understand about energy? Where does it come from? How does it flow across space from our sun? (Photo -Skylab 1973) What is the source of the Sun's energy? Why doesn't the Sun ever burn out? What is the source of the Sun's gravitational energy? How can it pull the Earth around in its orbit for trillions of years and not deplete its energy supply at all? Do we know the physical reasons why gravity attracts objects together instead of repelling them away from one another? Newton gave us a compelling description of this observation as an apparent attraction force, but provided no explanation for the existence and nature of this force itself. Does it really make sense that a force holds objects to the surface of planets, and moons in orbit, all with no known power source? Do we know if it is possible to create some type of anti-gravity device? What principles might underlie such a device? What principles underlie gravity itself? Men have been searching for these answers for thousands of years. Why should I believe that I could find the answers?

I checked out Albert Einstein. He found it necessary to continue searching for answers in spite of Newton's concept of gravity. Einstein arrived at a very different description of gravity, while the scientific community continues to search for still other explanations. Why do we have two separate explanations for the same effect of gravity in our science today? Why do we continue to search for still others? Is there something wrong here? Is there something we are not confronting? Why is our science so complex? Does our science get complex because we are not confronting the false data we have been given? Is there an ultimate simple explanation?

14) THE MECHANICS OF LIGHT

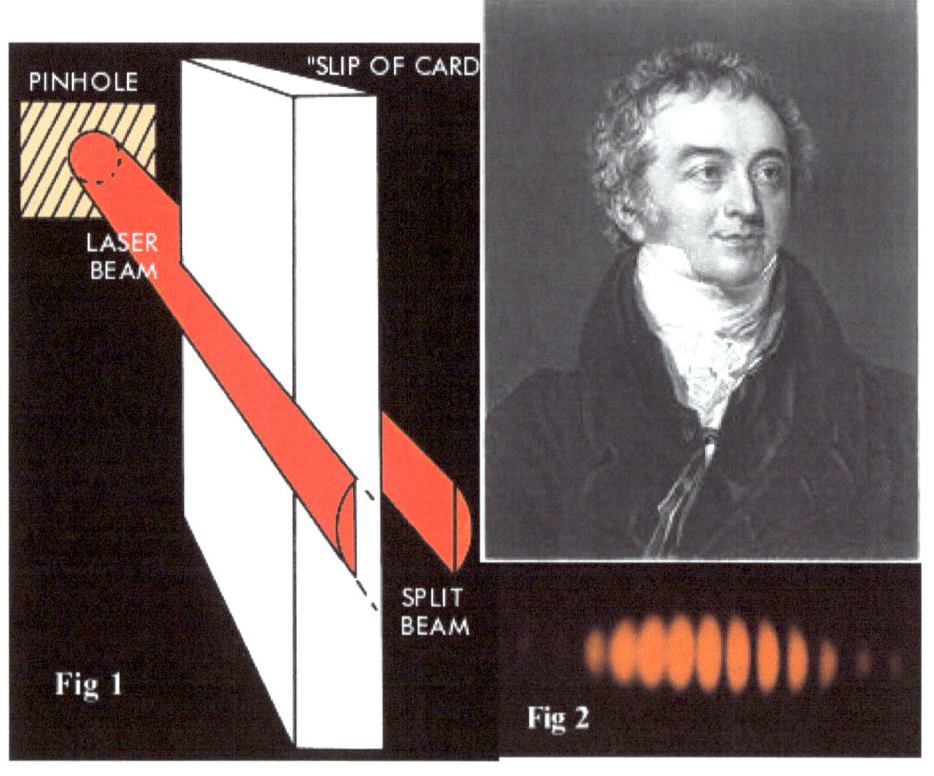

Do we truly understand light? Is it composed of waves or particles? And what is the ultimate source of this energy manifestation? Scientists today conceive of light as dualistic. It is both a wave and a particle, (a photon) depending on the experiment or situation. Thomas Young (left) was the first to demonstrate unequivocally that light is a wave. "The experiments I am about to relate ... may be repeated with great ease, whenever the sun shines, and without any other apparatus than is at hand to every one." This is how Thomas Young, speaking on November 24, 1803, to the Royal Society of London, began his description of the historic experiment. His audience, an august gathering of notables in science, was steeped in Isaac Newton's belief that light is made of tiny bullet-like particles, because it is always observed (or so Newton thought) to travel in straight beams, in contrast to the ripple-spreading behavior which Christian Huygens had linked with wave motion. "...It will not be denied by the most prejudiced," Young chided his skeptical listeners, "that the fringes [which are observed in Fig, 2] are produced by the interference of two portions of light." A narrow beam of sunlight was split with what Young described as "a slip of card, about one thirtieth of an inch in breadth (thickness)." The slip of card was held edgewise into the sunbeam, which was made to enter the room horizontally by means of a "looking glass" (mirror) and a tiny hole in a "window shutter". The sunbeam had a diameter slightly greater than the thickness of the card. When the card was placed properly it split the beam into two slivers, one passing on each side of the slip of card (Fig. 1). His talk was published in the following year's *Philosophical Transactions*, and was destined to become a classic, still reprinted and read today, giving in sparkling language the decisive evidence which first clearly demonstrated that light has the properties of waves. This is the wave-particle dualism in physics. Photons exhibit both the frequency dependent properties of waves and the mass, momentum, inertia properties of particles.

Today we have a theory known as quantum mechanics whose very creators readily describe as bizarre and mysterious. It casts doubt on our ability to fathom what is real and what is arbitrary.

Niels Bohr (left) - the man who mapped the atom said, "Anyone who is not shocked by quantum theory has not understood it." Werner Heisenberg, another pioneer of quantum mechanics, divided the universe into real and semi-real. He considered the elementary quantum particles to be only potentialities or possibilities, the semi-real. In this book I will present evidence that those potential particles are quantum waves created by beings just like you and me who live in the underlying thought universe, the place we go between lives on Earth. Once you conceive your relationship to your body, your mind, the thought universe and the Physical Universe, you will have achieved your immortality. Your own abilities are not nearly as limited as you have been made to believe. No one has ever permitted you to know how great you really are. Let me show you. You'll never fear death again.

15) THE MECHANICS OF MATTER

Where does matter come from? Why does it exhibit the qualities of gravity, inertia, dimensions, and chemistry? Did it all come from a singular "Big Bang" explosion of primordial superheated mass trillions of years ago to create our expanding universe which will eventually stop expanding due to gravitational forces and collapse upon itself as the "Big Crunch"? This Big Bang theory is currently believed by many of our top scientists to explain the origin of matter. Along with this theory goes the belief in black holes, gravitational forces holding planets in orbit and the apparently accelerating expansion of our universe.

Astronomers believe in the Big Bang creation theory today primarily due to the evidence that our universe is expanding. Most of the galaxies appear to be flying apart from one another as if from a great explosion. This belief is supported primarily by the fact that light arriving from distant galaxies is generally shifted to lower frequencies than expected. This is thought to indicate that all galaxies are speeding away from each other, much as the sound of a speeding train whistle drops in pitch once it passes and speeds away. This familiar drop in pitch is known by scientists as the Doppler Effect, named after Christian Doppler (1803-1853), and it is applied to light in order to explain the drop in frequency of light waves from distant galaxies. So, a lower-than-expected light frequency from a galaxy is interpreted by scientists to mean that it is speeding

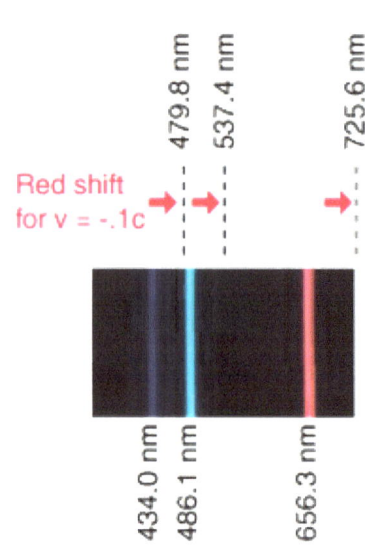

away from us. When energized atoms and molecules vibrate, they give off nearly mass-less light particles called photons. These photons travel as a wave, but because of quantum energy effects, a particular type of atom or molecule gives off only certain wavelengths of photon light. For example, when hydrogen atoms are giving off energy in the form of light, they emit light specifically at wavelengths of 410.2 nanometers (billionths of a meter), 434.0 nm, 486.1 nm, and 656.3 nm. This is called the emission spectra of hydrogen. Scientists can use the emission spectra of atoms and molecules to study the composition of stars. Scientists need to simply look very carefully at the intensity and wavelengths of the light given off by the star. A star containing hydrogen, for example, would have intense peaks of energy at 410.17, 434.05 nm, 486.13 nm, and 656.28 nm. These hydrogen emission peaks would be in addition to the ones associated with the other elements contained in the star.

In the 1920's, Edwin Hubble, while studying the stars of distant galaxies, found that for some, their emission spectra had peaks at 411.54 nm, 435.50 nm, 487.75 nm, and 658.47 nm. Hubble knew that these wavelengths did not correspond to any known element and that it was not likely that a combination of other elements or molecules was responsible. He did notice that these spectral lines corresponded to hydrogen's emission spectra except that they were all 0.0033 percent longer in wavelength than they should have been for a hydrogen spectrum. Hubble deduced that this red-shift must be because of a Doppler effect. His calculations showed that these galaxies must be moving away from earth at 1×10^6 m/s (one million meters per second)!

These measurements have convinced cosmologists that the universe is expanding. If you mentally reverse time by calculating the trajectory of these galaxies in the opposite direction, it becomes obvious that the expansion is the result of a point beginning. This is the basis of the "Big Bang" theory. But are these assumptions and interpretations correct? I say they are not. I have a simpler, more intuitive theory.

16) FAULTY MECHANICS OF THE BIG BANG THEORY

(1) Static universe models fit observational data better than expanding universe models.

Static universe models match most observations with no adjustable parameters. The Big Bang can match each of the critical observations, but only with adjustable parameters, one of which (the cosmic deceleration parameter) requires mutually exclusive values to match different tests. Without ad hoc theorizing, this point alone falsifies the Big Bang. Even if the discrepancy could be explained, Occam's razor favors the model with fewer adjustable parameters – the static universe model.

(2) The microwave "background" makes more sense as the limiting temperature of space heated by starlight than as the remnant of a fireball.

In 1926, Sir Arthur Eddington calculated the minimum temperature any body in space would cool to, given that it is immersed in the radiation of distant starlight. With no adjustable parameters, he obtained 3°K (later refined to 2.8°K, essentially the same as the observed, so-called "background", temperature. A similar calculation, although with less certain accuracy, applies to the limiting temperature of intergalactic space because of the radiation of galaxy light. So the intergalactic matter is like a "fog", and would therefore provide a simpler explanation for the microwave radiation, including its blackbody-shaped spectrum.

(3) Element abundance predictions using the Big Bang require too many adjustable parameters to make them work.

The universal abundances of most elements were predicted correctly by Hoyle in the context of the original Steady State cosmological model. This worked for all elements heavier than lithium. The Big Bang co-opted those results and concentrated on predicting the abundances of the light elements.

(4) The universe has too much large scale structure (interspersed "walls" and voids) to form in a time as short as 10-20 billion years.

The average speed of galaxies through space is a well-measured quantity. At those speeds, galaxies would require roughly the age of the universe to assemble into the largest structures (superclusters and walls) we see in space, and to clear all the voids between galaxy walls. But this assumes that the initial directions of motion are special, e.g., directed away from the centers of voids. To get around this problem, one must propose that galaxy speeds were initially much higher and have slowed due to some sort of "viscosity" of space. To form these structures by building up the needed motions through gravitational acceleration alone would take in excess of 100 billion years.

(5) The average luminosity of quasars must decrease with time in just the right way so that their average apparent brightness is the same at all redshifts, which is exceedingly unlikely.

According to the Big Bang theory, a quasar at a redshift of 1 is roughly ten times as far away as one at a redshift of 0.1. (The redshift-distance relation is not quite linear, but this is a fair approximation.) If the two quasars were intrinsically similar, the high redshift one would be about 100 times fainter because of the inverse square law. But it is, on average, of comparable apparent brightness. This must be explained as quasars "evolving" their intrinsic properties so that they get smaller and fainter as the universe evolves. That way, the quasar at redshift 1 can be intrinsically 100 times brighter than the one at 0.1, explaining why they appear (on average) to be comparably bright. It isn't as if the Big Bang has a reason why quasars should evolve in just this magical way. But that is required to explain the observations using the Big Bang interpretation of the redshift of quasars as a measure of cosmological distance.

(6) The ages of globular clusters appear older than the universe.

(7) The local streaming motions of galaxies are too high for a finite universe that is supposed to be everywhere uniform.

In the early 1990s, we learned that the average redshift for galaxies of a given brightness differs on opposite sides of the sky. The Big Bang interprets this as the existence of a puzzling group flow of galaxies relative to the microwave radiation.

(8) Invisible dark matter of an unknown but non-baryonic nature must be the dominant ingredient of the entire universe.

The Big Bang requires sprinkling galaxies, clusters, superclusters, and the universe with ever-increasing amounts of this invisible, not-yet-detected "dark matter" to keep the theory viable.

(9) The most distant galaxies in the Hubble Deep Field show insufficient evidence of evolution, with some of them having higher redshifts (z = 6-7) than the highest-redshift quasars.

The Big Bang requires that stars, quasars and galaxies in the early universe be "primitive", meaning mostly metal-free, because it requires many generations of supernovae to build up metal content in stars. But the latest evidence suggests lots of metal in the "earliest" quasars and galaxies.

(10) If the open universe we see today is extrapolated back near the beginning, the ratio of the actual density of matter in the universe to the critical density must differ from unity by just a part in 1059. Any larger deviation would result in a universe already collapsed on itself or already dissipated.

Anyone doubting the Big Bang in its present form (which includes most astronomy-interested people outside the field of astronomy, according to one recent survey) would have good cause for that opinion and could easily defend such a position. This is a fundamentally different matter than proving the Big Bang did not happen, which would be proving a negative – something that is normally impossible. (E.g., we cannot prove that Santa Claus does not exist.) The Big Bang, much like the Santa Claus hypothesis, no longer makes testable predictions wherein proponents agree that a failure would falsify the hypothesis. Instead, the theory is continually amended to account for all new, unexpected discoveries. Indeed, many young scientists now think of this as a normal process in science! They forget or were never taught that a model has value only when it can predict new things that differentiate the model from chance and from other models before the new things are discovered. Explanations of new things are supposed to flow from the basic theory itself with at most an adjustable parameter or two, and not from add-on bits of new theory.

17) THE MECHANICS OF REALITY DISTORTION

There are many misperceptions we humans make that distort reality for us and throw us off the trail of discovery of truth. For instance, humans think that they <u>are</u> their bodies and don't distinguish between themselves, spiritual beings, and their bodies. The very fact that they can close their eyes and see things in their mind should be enough to show them that they are spiritual beings. They create things with their imagination, with zero-point energy. But they think that their brain is their mind. They think that they control their body with chemical nerve impulses. They think that God created the universe instead of them. They think that God is omnipotent and they are powerless. They think that matter is solid when in fact there is no discrete particle or solid chunk of anything in the whole Physical Universe! It is all standing waves; agitation within agitation. That is why the Periodic Table of the Elements is arranged in harmonic columns. In fact, there is no matter; it is all energy of vibration held together by electrical charge.

Living in a meat body really throws us off. As a human, the physical laws of the underlying quantum field are counter intuitive. We interpret manifestations in the physical universe according to our perceptions as bodies, not according to the actual physics involved in our lives. For instance, we think that the photons coming from a bright light are stronger, have more energy than photons of the same color coming from a dim light. That is intuitive for us. The truth is that any photon of a given color has exactly the same energy as any other photon of that color. The difference in brightness that we see is due solely to the number of photons given off by that light.

The brighter light has more photons being given off. Each photon has an energy, $E=hf$. The energy is Plank's constant (h) times the frequency.

Another thing we don't appreciate is that all energy interactions in the Physical Universe come in discrete packets or quanta. When you heat up the water in your tea kettle on the stove, it is intuitive that there is a smooth and gradual flow of heart from the gas fire into the kettle. But such is not the case. In reality tiny packets or quanta of energy are being put into the water but they are so small that you don't notice each little jump of energy. With so many atoms and water molecules involved, it appears as if the water temperature smoothly rises until it is boiling. The effect can be compared to a digital photo of a car that looks like a true copy of the car until you put the photo under a microscope and discover that it is made of tiny discrete dots called pixels.

The Physical Universe can be viewed from three perspectives, or range of sizes. There is the Quantum Realm, the Human Realm and the Cosmos Realm. Our human senses have limitations. It is difficult for us to observe things around us that are very large or very small. We are most familiar with the human realm. It encompasses lengths, speeds and masses which we can sense with our eyes ears and muscles. This range can be extended with microscopes, telescopes and other instruments.

The second range is the quantum realm, which is smaller than the human realm. Science has found that physics is different there; it is dominated by the laws of quantum theory. In the quantum realm we are dealing with one particle at a time which looks like a wave. The most intriguing realization of all to come from modern physics is the fact that there is some sort of connection between quanta that completely ignores space-time separation, even if the quanta are at opposite ends of the Universe. These are called **nonlocal connections** and while being dealt with in various ways in theory, they remain a completely illogical yet entirely real aspect of the quantum world. Somehow, even though we see macroscopic matter as separated by space, we must consider the quanta that make up this matter as existing in a dimension in which *nothing is separate*. Exactly how this can be certainly defies objective logic, but it is real -- it is as fundamental to physical reality as the quanta are themselves. The *experience of matter-energy in space-time* requires a level of understanding that goes well beyond objective logic.

COSMIC MECHANICS

The third range is the cosmos realm which includes things larger than the human realm. It includes the speed of light (3×10^8 meters per second) which is the greatest possible speed of any object. Three new phenomena are involved: Special Relativity, General Relativity, and the Hubble Rule. This realm is characterized by changing mass, length, and time when things get to moving close to the speed of light. Underlying these three realms is the zero-point field or thought universe. The lowest velocity that exists in the zero-point field is the speed of light, c. The highest velocity that exists in the Physical Universe is the speed of light, c.

Here is an illustration of the three realms, quantum, human and cosmos.

18) THE MECHANICS OF GRAVITATION

Newton stated that there is an attracting force in nature emanating from all objects, pulling them toward one another with a strength that increases with their masses and decreases with the distance between them squared. His formula for the force between masses is; $F = G * (m_1 m_2)/R^2$ where m_1 and m_2 are the masses of the two objects and R is the distance (radius) between their centers. G is a constant, called the gravitational constant.

When Newton introduced this concept, it ushered a completely new force of nature into our awareness and our science. It was not merely an abstract model of observations, but a statement of an actual force in nature emanating from objects – varying in strength with their mass, which we can lift, and their distance, which we can measure. This is a concept that we are now taught as children and have grown accustomed to. But where did this revelation come from? Somehow we went from a vague suspicion that an attracting force might be operating in the world around us, to a definite statement of its existence, its source in all material objects, and its precise behavior captured in an equation. How did this happen? I will show that Newton's gravitational theory is actually a completely superfluous and unnecessary invention that is based on a logically and scientifically flawed assumption. It is one of the lies that keep the Physical Universe solid; because if we could see our universe as it is, it would lose its power and disappear. Even Newton admitted that he was making it up as he went along. That a force of attraction might exist between two distant objects, he once wrote in a letter, is "so great an Absurdity that I believe no Man who has in philosophical Matters a competent Faculty of thinking can ever fall into it." Yet fall into it we all do on a daily basis, and physicists are no exception.

The discovery of the true nature of gravity would presumably not be without philosophical consequences of the civilization-altering variety. Cosmologists often refer to this possibility as "the ultimate Copernican revolution": not only are we not at the center of anything; we don't even know what's out there. "We're just a bit of pollution," Lawrence M. Krauss, a theorist at Case Western Reserve, said in 2006 at a public panel on cosmology in Chicago. "If you got rid of us, and all the stars and all the galaxies and all the planets and all the aliens and everybody, then the universe would be largely the same. We're completely irrelevant." Science is full of Homo sapiens-humbling insights. But the trade-off for these lessons in insignificance has always been that at least now we would have a deeper — simpler — understanding of the universe. That the more we could observe, the more we would know. But what about the less we could observe? What happens to new knowledge then? It's a question cosmologists have been asking themselves lately, and it might well be a question we'll all be asking ourselves soon, because if they're right, then the time has come to rethink a fundamental assumption: When we look up at the night sky, we're seeing the universe. Right?

Not so. Not even close and I'll tell you why. What we perceive is what we have created. From the photons that reach our eyes we create a magnificent universe in present time of stars and constellations that existed 5 years to 100 million years ago. Some of those stars aren't there any more. They have burned out or exploded in a super nova. And those are only the stars in our own little neighborhood. There are stars and universes without number beyond the reach of our perception.

MICROWAVE BACKGROUND RADIATION

In 1963, two scientists at Bell Labs in New Jersey discovered a microwave signal that came from every direction of the heavens. Theorists at nearby Princeton University soon realized that this signal might be the echo from the beginning of the universe, as predicted by the big-bang hypothesis. Take the idea of a cosmos born in a primordial fireball and cooling down ever since, apply the discovery of a microwave signal with a temperature that corresponded precisely to the one that was predicted by theorists — 2.7 degrees above absolute zero — and you have the universe as we know it. Not Newton's universe, with its stately, eternal procession of benign objects, but Einstein's universe, violent, evolving, full of births and deaths, with the grandest birth and, maybe, death belonging to the cosmos itself.

But then, **in the 1970s**, astronomers began noticing something that didn't seem to fit with the laws of physics. They found that spiral galaxies like our own Milky Way were spinning so fast that they should have long ago wobbled out of control, shredding apart, shedding stars in every direction. Yet clearly they had done no such thing. They were living fast but not dying young. This seeming paradox led theorists to wonder if a halo of a hypothetical something else might be cocooning each galaxy, dwarfing each flat spiral disk of stars and gas at just the right mass ratio to keep it gravitationally intact. The observed effects of gravity in the Physical Universe did not match with Newton's or Einstein's theories and mathematics. Early in 1998, two teams of prominent astronomers had amassed data from more than 50 supernovae between them — data that would reveal yet another oddity in the cosmos; the two teams announced that they had each independently reached the same conclusion, and it was the opposite of what either of them expected. The rate of the expansion of the universe was not slowing down. Instead, it seemed to be speeding up. It's the most profound mystery in all of science.

This is a dramatic revolution in understanding. It's a sudden liberation from old limits. Each of these discoveries is properly described as a 'paradigm shift." A paradigm is a scheme for understanding and explaining certain aspects of reality. A paradigm shift is a distinctly new way of thinking about old problems ... the problem is that you can't embrace the new paradigm unless you let go of the old, and that can be very difficult.

19) DO PLANETS ACTUALLY TUG ON PASSING BODIES?

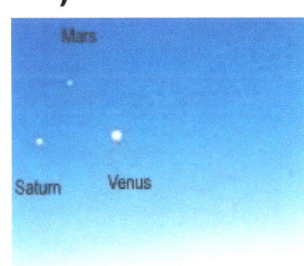

Planets don't actually exert a tugging force on passing bodies. At left is a view of Mars, Saturn, Venus and Mercury during the conjunction of May 3, 2002. This conjunction gave rise to no unusual tidal or volcanic effects. Our current belief in the remote gravitational tug from distant planets gives rise to periodic predictions of unusual increases in tidal effects on Earth. These predictions arise from an expected increase in the "gravitational pull" upon the Earth when a number of planets are due to align with the Earth in their orbits around the sun; yet, the predicted effects never seem to materialize. The reason nothing happens during these conjunctions, of course, is because there is no gravitational force emanating from these planets to affect the Earth. As Mach's principle states, *Every local inertial frame is determined by the composite matter of the universe.* It isn't done by just the local planets.

20) THE PIONEER ANOMALLY

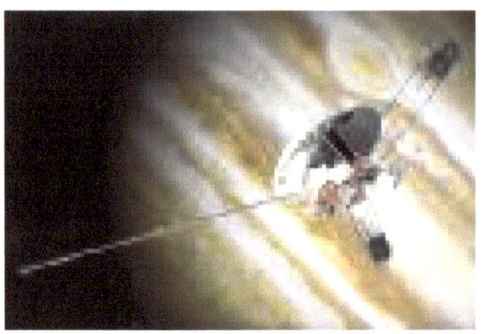

The Pioneer 10 and 11 spacecraft exhibited an unexplained "Gravitational Anomaly" as they sped past Pluto and continued out of our solar system. NASA scientists have noted and published observations of an apparent and unexplained gravitational pull on both spacecraft back toward the sun. This pull greatly exceeds their expectations of gravity at that distance. Newton's law of gravity states that the gravitational attractive force diminishes with the square of the distance between the masses. So when the distance doubles, the attractive force is one-quarter and when the distance is ten times as far, the attractive force is one-hundredth. So how much "gravitational force" is expected to be exerted on a spacecraft three and a half billion miles from the sun? It would be practically negligible, yet NASA is reporting significant reductions in velocity of the spacecraft as they head out of our solar system. Attempts to explain this effect using all known or even proposed gravitational or physical theories have so far been unsuccessful.

However, when we look at this mystery from the perspective of *the wave structure of matter*, discovered by Dr. Milo Wolff, the journeys of these spacecraft take on a very different quality. The situation now changes from that of two spacecraft being pulled back by an unexplained "gravitational force," to that of space resonances being relocated by the minimum amplitude principle. The Pioneer spacecraft are not actually feeling the tug of the sun's "gravity" at all – and they never did all throughout their journeys across the solar system since such "gravitational forces" do not exist in *the wave structure of matter*. Inertia and gravity are caused by the interaction of the space resonances with the zero-point longitudinal waves.

PRINCIPLE OF EQUIVALENCE: We know that the effects of gravity and acceleration cannot be distinguished. This is known, by our scientists, as the "Principle of Equivalence." If you were

accelerated upward in an elevator in space at 9.8 meters per second per second (this is not a typographical error and I'm not stuttering - it is 9.8 m/sec^2) you would experience your normal weight on Earth. Also, if you were on the inside of the wheel (not the hub) in a space station like this in the film *2001, A Space Odyssey*, you would experience normal gravity if the wheel turned in space to change the direction of your inertia by 9.8 meters per second per second. Similarly, gravity is accelerating us at 9.8m/sec^2 toward the center of the earth.

21) MECHANICS OF A HELIUM BALLOON

A helium balloon inside a closed car drifts toward the direction in which the car is turning. This is a very upsetting phenomenon the first time you see it. You would expect to see the balloon thrown to the outside of the turn, but just the opposite occurs. A plum bob swings to the outside of the turn, but the balloon does just the opposite.

Common experience would tell us that an object floating free should be thrown toward the outside of a curve. The reason that a helium balloon does not is because helium, being less massive, accelerates (or decelerates) at a slower rate than the air around it. If you have a hard time following that, so do I. I felt strongly that any possible effect would be so small as to be unobservable in a car, so I tested the theory.

I tied the balloon down to the center of the car floor so that it was free to float in any horizontal direction, and drove off accelerating and braking and turning sharp corners. I also had a plum bob hanging from the ceiling over the balloon. It was positively eerie! A lifetime of experience seemed to tell me that this phenomenon was wrong. Any acceleration caused the bob and the balloon to move in opposite directions. If you slam on the brakes, things on the back seat fly forward but the balloon jerks backward. If you accelerate, inertia tugs backward on your head, but the balloon seems to be racing to get ahead of the car.

Gravity and acceleration have exactly the same effect (in fact, they are identical). Assume you were standing in a rocket ship someplace in the galaxy, and that the rocket ship was accelerating at the rate of 9.8 meters per second per second. If your head was pointed in the direction of travel, you would not be able to tell the difference between the push on the bottom of your feet and the normal gravity you would feel on earth. Sitting in a car on earth, you feel the acceleration due to gravity on the seat of your pants, except when you add an additional component of acceleration or deceleration by stepping on the throttle or the brake. Strictly speaking, there is no such thing as "deceleration"--it's just acceleration backward in the direction that you've already traveled. In a turn, you are accelerated sideways. When you combine the downward component of gravity with the sideways (or forward, or backward) component due to acceleration, the total "field" is briefly tilted, as the fluid level in a car partially filled with water would show you. Physicists call this a superposition of fields. A plumb bob would point perpendicular to the surface of the water, and when there is only air in the car, the air is pulled "down" in the direction in which the plumb bob points. The helium balloon thinks so, too, and like helium balloons should, it tries to move "up" against this temporarily tilted artificial gravity field. One way to mentally picture what is happening is to think that the balloon is trying to move toward air of lesser density, and that when a car is accelerating (sideways, backward or forward), the air inside is briefly rarefied on the side (or end) toward the direction in which it is accelerating, and compacted against the opposite wall.

In **1893 Ernst Mach** studied the rotation of the Sun, planets, Moon and stars and came up with the startling conclusion that Newton's law of inertia depends upon the distant stars. Mach reasoned that there must be a causal connection between the existence of all the distant matter in the universe and any local inertial reference frame. Therefore, he postulated what is now known as Mach's Principle: ***Every local inertial frame is determined by the composite matter of the universe.*** The principle implies that the magnitude and direction of inertial forces on a local mass is determined by relative motion with respect to the total mass of the universe. That's like saying that when the subway jerks, you are thrown down by the mass of the distant stars! Now I was starting to get the big picture about my relationship to the Physical Universe.

So now you know why spiral galaxies don't fling their outer stars into the cosmos. **The orbits of a galaxy's stars are established by the total mass of the universe, not just by the mass of the center of the local spinning galaxy.**

Gravity and acceleration are equivalent effects. Is gravity a pulling force or a pushing force? When a brick drops, is it a pulling force or a pushing force that accelerates it toward the ground? When a balloon rises, is it a pushing force or a pulling force that lifts it? With a balloon, the laws of buoyancy are used to calculate the lift. When the weight of the displaced air is greater than the weight of the helium and balloon, it will rise. You don't have to pray to the lift god to experience this effect the same as you don't have to pray to God in Heaven to keep the planets in their orbits. It is all Physics, the laws of energy interchanges.

COSMIC MECHANICS

In the case of buoyancy, the balloon rises and expands in the atmosphere until the weight of the balloon equals the weight of the displaced air. At higher altitudes the air is thinner, less dense, so there is a point where the balloon ceases to rise because it weighs as much as the displaced air weighs at that altitude.

The same is true of falling bodies such as artillery shells and satellites. There is a balance of forces. For instance, the centripetal force of gravity pulling the Moon toward the Earth is exactly balanced by the centrifugal force of the speeding Moon's mass pulling it out of orbit. Notice that the speeding Moon-Universe mass (inertial energy) is pulling it out of orbit while the Earth-Moon mass (gravitational energy) is pulling it back into orbit.

We can get the effect of anti-gravity by increasing the buoyancy or get the effect of gravity by decreasing buoyancy. With falling bodies, we can get the effect of anti-gravity by increasing the velocity of a satellite and we can get the effect of gravity by decreasing the velocity. Varying the velocity will move the satellite to higher or lower orbits.

We experience two broad classes of energy in the Physical Universe. The first is the energy placed in the zero-point field by cosmic beings. This is postulated or thought energy. It is eternal and doesn't diminish over time. Its source is the cosmic beings and no battery or dynamo or fire is needed for its manufacture. This is magnetic, electrostatic and gravitic energy. Unfortunately we cannot get work out of this form of energy. The second class of energy is the cause and effect energy in the chaos of matter which cascades down like dominoes from higher quality energy to lower as entropy increases. An example is light energy as it evaporates water from the ocean which falls into a river which feeds a dam which generates electricity which goes to an electric drill which drills a hole in wood and the resultant heat dissipates into the environment. This is the cascading effect of energy traveling from a high order (light) through various electro-mechanical-chemical processes to a low order (heat). The laws of thermodynamics apply to this class of energy. They are the **laws of conservation of energy** and the degeneration of energy from higher to lower quality forms. These laws state that energy can not be created nor destroyed; it can only be changed from one form to another, such as when electrical energy is changed into heat energy. In Thermodynamics these laws are: **O)** If two thermodynamic systems are in thermal equilibrium with a third, they are also in thermal equilibrium with each other. **1)** *In any process,* *the total energy of the universe remains constant.* **2)** *In an isolated system, concentrated energy disperses over time, and consequently less concentrated energy is available to do useful work. Energy dispersal also means that differences in temperature, pressure, and density even out. Again roughly speaking, thermodynamic entropy is a measure of energy dispersal.* **3)** *As temperature approaches absolute zero, the entropy (the amount that localized energy in a system is dispersed or spread out) of a system approaches a constant. In other words, near absolute zero, there is bits of energy lying around but not enough to cause any flows or motion among the particles.*

Apathetic scientific materialism presents a dim view of these laws. It says **0)** You have to play the game; **1)** You can't win; **2)** You can't break even, except on a very cold day; and **3)** It doesn't get that cold.

22) THE MECHANICS OF CONSERVATION OF ENERGY

Our scientists and professors of physics have told us that matter and energy cannot be created nor destroyed. This is today's stable datum for Physics and Chemistry. But I know that I can create space and time because I created the little 21 gram orb that I carry with me from lifetime to

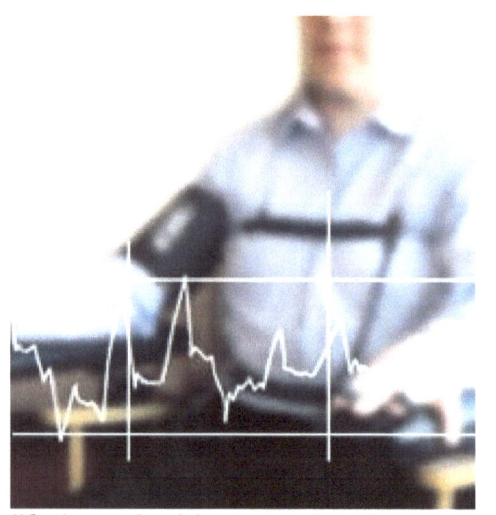
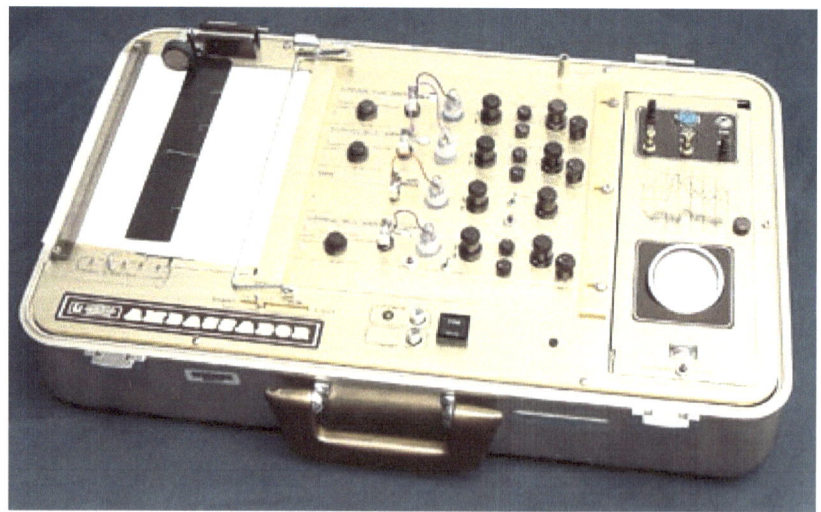

lifetime. And I can probably create a heck of a lot more than that! I've been taking that orb with me for two hundred trillion years and I'm sick of it (literally, it causes disease in my body). When the mental mass, or orb, is outside the body it can be photographed by a camera with a CCD sensitive to infrared frequencies. The term **orb** is the popular name given to typically circular anomalies appearing in photographs. In the Spanish language they're commonly referred to as *canoplas*. In photography and video, orbs appear to be balls or smears of light with an apparent size in the image ranging from a golfball to

COSMIC MECHANICS

a basketball. Orbs sometimes appear to be in motion, leaving a trail behind them. The trail is caused by the flash dimming out during its 1/1,000 second burst. If the orb is rising quickly the tail will be pointed up but our intuition would tell us that the orb is moving downward. That's because the orb is captured in the CCD initially when the bright flash goes off and then as the capacitor discharges through the flash tube the brightness diminishes as the orb gains altitude – leaving a rising trail. A cosmic being in distress caused by a recent death due to drowning or accident will show as a small dense orb. It is small, opaque, and very bright in the flash of the camera. It has lots of mass and therefore weight and energy. After the being chills out its orb will be larger, transparent, less dense, dim and gossamer. It will NOT have lost mass, weight OR energy. It just spreads out, gets less dense. This correlates with psycho-galvanometer readings. A subject's skin resistance will be high when the orb is condensed. When the subject relaxes and the orb expands and thins out the skin resistance will drop considerably. This also correlates with biophotonic inter-cellular communication. When the density of the orb is high during times of stress and pain, the biophotonic communication coherence will be low, analogous to a crying baby. But when the density of the orb is low in its expanded, relaxed, healthful state, the coherence of the photonic communication between cells will be very high, analogous to a cooing baby or a person listening to a symphony by Mozzart.

Let's now examine the mechanics of waves since they are the substance of the Physical Universe. Everything in the Physical Universe is composed of waves. There is absolutely nothing that is solid.

23) WAVE MECHANICS

"Quantum mechanics long ago destroyed materialism for all time, but it just hasn't percolated through the prevailing scientific dogma yet". –Tom Bearden

To understand yourself and your relationship to your mind, your brain, your body and the Physical Universe, you must first understand waves. So we are going to review wave mechanics. This is all single spaced so if you are not technically minded, you can skim over it.

A rock tossed into the water will create a circular disturbance which travels outwards in all directions.

Everything is composed of waves. The Physical Universe is composed of waves. There is not one particle in it. No matter how deep physicists peer into the depths of atoms, they cannot find anything solid. It is just agitation within agitation; waves. It is difficult for humans to understand this principle because everything seems so solid to them. They think that they are a body and that they think with their brains. They cannot see, feel and touch that they are spiritual beings, a life static that is composed of nothing, no matter, energy, space, or time. But they can create these things. This is the greatest suppressed secret of all time. **Humans are not their bodies and no one has ever allowed them to know how great they really are. They are matter, energy, space and time creation units!** They create things with waves. The best things in life aren't things; they are nothings! They are us.

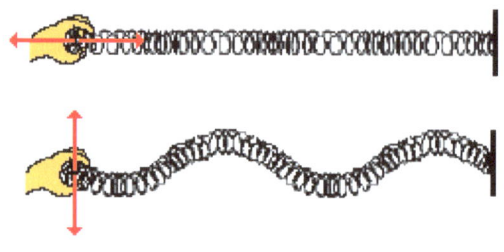

Slinky waves can be made by vibrating the first coil back and forth in either a horizontal or a vertical direction.

There are a few clues lying around that remind us of the life we had before assuming a body in the Physical Universe. For example, we are mystified with magnets. Where does their mysterious force come from and why does their energy never run out? Why haven't scientists discovered the source of the mysterious "gravity wave" and why does gravity continue to pull the Earth around the Sun in orbit forever without running out of energy? What holds the molecules of a swing chain together so tight that a 300 pound man can swing forever without fear of falling when the chain breaks? Some forms of energy, like gasoline, food, electricity and heat have to be replenished. You would get very tired and hungry turning a hand crank generator to keep a small lamp lit. But the Sun belches out radiant energy forever with no sign of ever running down. What is the source of this eternal energy? Sound waves, visible light waves, radio waves, microwaves, water waves, sine waves, cosine waves, telephone cord waves, earthquake waves, waves on a string, and slinky waves are just a few of the examples of waves that we can see in our environment. We study the physics of waves because it provides a rich glimpse into the physical world which we seek to understand and of the thought universe, which underlies everything.

If a slinky is stretched out from end to end, a wave can be introduced into the slinky by either vibrating the first coil up and down vertically or back and forth horizontally. A wave will subsequently be seen traveling from one end of the slinky to the other. As the wave moves along the slinky, each individual coil is seen to move out of place and then return to its original position. The coils always move in the same direction that the first coil was vibrated. A continued vibration of the first coil results in a continued back and forth motion of the other coils. If looked at closely, one notices that the wave does not stop when it reaches the end of the slinky; rather it seems to bounce off the end and head back from where it started. A slinky wave provides an excellent mental picture of a wave. The shape of the wave is influenced by the frequency at which it is vibrated. If the student vibrates the chord rather frequently, then a short wave is created; and if the student vibrates the chord at a low frequency (not so often), then a long wave is created. The energy in the wave is proportional to its frequency. The person demonstrating the wave will notice that a gentle and slow arm motion will create a low frequency wave that takes no effort to produce and maintain. But the demonstrator's arm will quickly tire if he vibrates the slinky fast enough to produce lots of short waves in the slinky. This principle is expressed in the equation, $E=hf$, which shows that the energy of a vibrating system is proportional (h is a constant) to the frequency.

You may have encountered waves in Math class in the form of the sine and cosine function. We often plotted $y=B\sin(Ax)$ on our calculator or by hand and observed that its graphical shape resembled the characteristic shape of a wave. There was a crest and a trough and a repeating pattern. If we changed the constant A in the equation, we noticed that we could change the *length* of the wave. And if we changed B in the equation, we noticed that we changed the *height* of the wave. In math class, we encountered the underlying mathematical functions which describe the physical nature of waves.

Finally, we are familiar with microwaves and visible light waves. While we have never seen them, we believe that they exist because we have witnessed how they carry energy from one location to another. And similarly, we are familiar with radio waves and sound waves. Like microwaves, we have never seen them. Yet we believe they exist because we have witnessed the signals which they carry from one location to another and we have even learned how to tune into those signals through use of our ears or a tuner on a television or radio. Waves, as we will learn, carry energy from one location to another. And if the frequency of those waves can be changed, then we can also carry a complex signal which is capable of transmitting an idea or thought from one location to another.

A medium can be modeled by a series of particles connected by springs. As one particle is displaced, ... the spring attaching it to the next is stretched and begins to exert a force on its neighbor, thus displacing the neighbor from its rest position.

What is a Wave?

But what makes a wave *a wave*? What characteristics, properties, or behaviors are shared by the entire phenomenon which we typically characterize as being a wave? How can waves be described in a manner that allows us to understand their basic nature and qualities?

A wave can be described as a disturbance that travels through a medium from one location to another location. Consider a slinky wave as an example of a wave. When the slinky is stretched from end to end and is held at rest, it assumes a natural position known as the **equilibrium or rest position.** The coils of the slinky naturally assume this position, spaced equally far apart. To introduce a wave into the slinky, the first particle is displaced or moved from its equilibrium or rest position. The particle might be moved upwards or downwards, forwards or backwards; but once moved, it is returned to its original equilibrium or rest position. The act of moving the first coil of the slinky in a given direction and then returning it to its equilibrium position creates a **disturbance** in the slinky. We can then observe this disturbance moving through the slinky from one end to the other. If the first coil of the slinky is given a single back-and-forth vibration, then we call the observed motion of the disturbance through the slinky a *slinky pulse*. A **pulse** is a single disturbance moving through a medium from one location to another location. However, if the first coil of the slinky is continuously and periodically vibrated in a back-and-forth manner, we would observe a repeating disturbance moving within the slinky which endures over some prolonged period of time. The repeating and periodic disturbance which moves through a medium from one location to another is referred to as a **wave**.

When a slinky is stretched, the individual coil assume an equilibrium or rest position.

When the first coil of the slinky is repeatedly vibrated back and forth, a disturbance is created which travels through the slinky from one end to the other.

But what is meant by the word *medium*? A **medium** is a substance or material which carries the wave. You have perhaps heard of the phrase *news medium*. The news media refers to the various institutions (newspaper offices, television stations, radio stations, etc.) within our society which carry the news from one location to another. The news moves through the media. The media doesn't make the news and the media isn't the same as the news. The news media is merely the *thing* that carries the news from its source to various locations. In a similar manner, a wave medium is the substance which carries a wave (or disturbance) from one location to another. The wave medium is not the wave and it doesn't make the wave; it merely carries or transports the wave from its source to other locations. In the case of our slinky wave, the medium through which the wave travels is the slinky coils. In the case of a water wave in the ocean, the medium through which the wave travels is the ocean water. In the case of a sound wave moving from the church choir to the pews, the medium through which the sound wave travels is the air in the room. The medium through which scalar zero-point waves travel is the thought, or spiritual universe, the place we visit between earth lives, which is our home.

To fully understand the nature of a wave, it is important to consider the medium as a series of interconnected or merely interacting *parts*. In other words, the medium is composed of parts which are capable of interacting with each other. The interactions of one part of the medium with the next adjacent part allow the disturbance to travel through the medium. In the case of the slinky wave, the *parts* or interacting parts of the medium are the individual coils of the slinky. In the case of a sound wave in air, the *parts* or interacting parts of the medium are the individual molecules of air.

Consider the presence of a wave in a slinky. The first coil becomes disturbed and begins to push or pull on the second coil; this push or pull on the second coil will displace the second coil from its equilibrium position. As the second coil becomes displaced, it begins to push or pull on the third coil; the push or pull on the third coil displaces it from its equilibrium position. As the third coil becomes displaced, it begins to push or pull on the fourth coil. This process continues in consecutive fashion, each individual *particle* acting to displace the adjacent particle; subsequently the disturbance travels through the medium. The medium can be pictured as a series of particles connected by springs. As one particle moves, the spring connecting it to the next particle begins to stretch and apply a force to its adjacent neighbor. As this neighbor begins to move, the spring attaching the neighbor to its neighbor begins to stretch and apply a force on its adjacent neighbor.

When a wave is present in a medium (that is, when there is a disturbance moving through a medium), the individual particles of the medium are only temporarily displaced from their rest position. There is always a force acting upon the particles which restores them to their original position. In a slinky wave, each coil of the slinky always returns to its original position. In a water wave, each molecule of the water always returns to its original position. It is for this reason, that a wave is said to involve the movement of a disturbance without the movement of matter. This is equally true in the Physical Universe as it is in the thought universe.

Waves are said to be an **energy transport phenomenon**. As a disturbance moves through a medium from one part to its adjacent part, energy is being transported from one end of the medium to the other. In a slinky wave, a person imparts energy to the first coil by doing work upon it. The first coil receives a large amount of energy which it subsequently transfers to the second coil. When the first coil returns to its original position, it possesses the same amount of energy as it had before it was displaced. The first coil transferred its energy to the second coil. The second coil then has a large amount of energy which it subsequently transfers to the third coil. When the third coil returns to its original position, it possesses the same amount of energy as it had before it was displaced. The third coil has received the energy of the second coil. This process of energy transfer continues as each coil interacts with its neighbor. In this manner, energy is transported from one end of the slinky to the other, from its source to another location. The same thing happens in the zero-point thought universe. A being, devoid of matter, energy, space, time or location, causes a wave disturbance in its all-pervasive, absolutely still, space. This disturbance travels at super-luminal velocities in all directions from the source carrying the information or thought of the originator. Fantastic amounts of energy can be transmitted in this way which greatly exceeds the energy of the brightest supernova our astronomers have ever seen.

This characteristic of a wave as an energy transport phenomenon distinguishes waves from other types of phenomenon. Consider a common phenomenon observed at a softball game - the collision of a bat with a ball. A batter is able to transport energy from her to the softball by means of a bat. The batter applies a force to the bat, thus imparting energy to the bat in the form of kinetic energy. The bat then carries this energy to the softball and transports the energy to the softball upon collision. In this example, a bat is used to transport energy from the player to the softball. However, unlike wave phenomena, this phenomenon involves the transport of matter. The bat must move from its starting location to the contact location in order to transport energy. In a wave phenomenon, energy can move from one location to another, yet matter does not move. A wave transports its energy without transporting matter. This is an important principle, especially when considering the transport of zero-point thought energy where there is no matter, just postulated waves created by the imagination of spiritual beings.

Waves are seen to move through an ocean or lake; yet the water always returns to its rest position. Energy is transported through the medium, yet the water molecules are not transported. A magnetic force is exerted on a piece of iron, yet no matter is moving. In conclusion, a wave can be described as a disturbance which travels through a medium, transporting energy from one location (its source) to another location without transporting matter. Each individual wave structure of the medium is temporarily displaced and then returns to its original equilibrium position. **In the thought universe, a disturbance ripples through the stillness carrying energy and information through space.**

CATEGORIES OF WAVES

Waves come in many shapes and forms. While all waves share some basic characteristic properties and behaviors, some waves can be distinguished from others based on some very observable (and some non-observable) characteristics. It is common to categorize waves based on these distinguishing characteristics.

One way to categorize waves is on the basis of the direction of movement of the individual particles of the medium relative to the direction which the waves travel. Categorizing waves on this basis leads to three notable categories: transverse waves, longitudinal or scalar waves, and surface waves.

A **transverse wave** is a wave in which particles of the medium move in a direction perpendicular to the direction which the wave moves. If a slinky is stretched out in a horizontal direction across the classroom, and a pulse is introduced into the slinky on the left end by vibrating the first coil up and down, then energy will begin to be transported through the slinky from left to right. As the energy is transported from left to right, the individual coils of the medium will be displaced upwards and downwards. In this case, the particles of the medium move perpendicular to the direction in which the pulse moves. This type of wave is a transverse wave. Transverse waves are always characterized by particle motion being perpendicular to wave motion.

Dr. F. Lee Aeilts

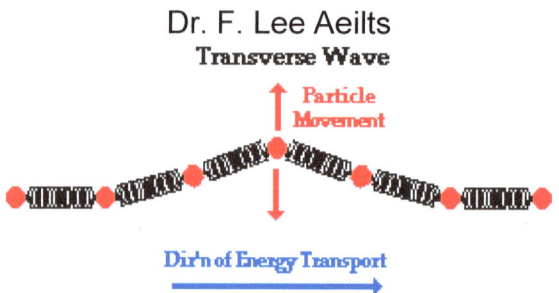

A **longitudinal wave** is a wave in which particles of the medium move in a direction <u>parallel</u> to the direction which the wave moves. If a slinky is stretched out in a horizontal direction across the classroom, and a pulse is introduced into the slinky on the left end by vibrating the first coil left and right, then energy will begin to be transported through the slinky from left to right. As the energy is transported from left to right, the individual coils of the medium will be displaced leftwards and rightwards. In this case, the particles of the medium move parallel to the direction which the pulse moves. This type of wave is a longitudinal wave. Longitudinal waves are always characterized by particle motion being <u>parallel</u> to wave motion.

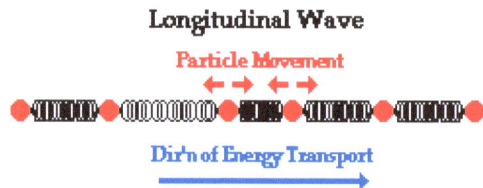

A sound wave is a classic example of a longitudinal wave. As a sound wave moves from the lips of a speaker to the ear of a listener, particles of air vibrate back and forth in the same direction and the opposite direction of energy transport. Each individual particle pushes on its neighboring particle so as to push it forward. The *collision* of particle #1 with its neighbor serves to restore particle #1 to its original position and displace particle #2 in a forwards direction. This back and forth motion of particles in the direction of energy transport creates regions within the medium where the particles are pressed together and other regions where the particles are spread apart. Longitudinal waves can always be quickly identified by the presence of such regions. This process continues along the *chain* of particles until the sound wave reaches the ear of the listener.

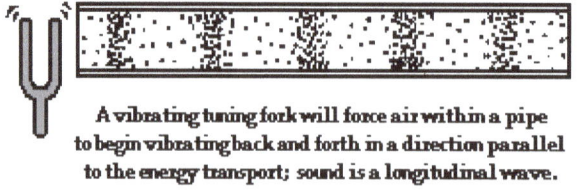

Waves traveling through a solid medium can be either transverse waves or longitudinal waves. Yet waves traveling through the bulk of a fluid (such as a liquid or a gas) are always longitudinal waves. **Waves traveling in the zero-point field are longitudinal waves; also known as scalar.** They are not like radio and light waves which are transverse waves. Transverse waves require a relatively rigid medium in order to transmit their energy. As one particle begins to move it must be able to exert a pull on its nearest neighbor. If the medium is not rigid as is the case with fluids, the particles will slide past each other. This sliding action which is characteristic of liquids and gases prevents one particle from displacing its neighbor in a direction perpendicular to the energy transport. It is for this reason that only longitudinal waves are observed moving through the bulk of liquids such as our oceans.

Any wave moving through a medium has a source. **In the zero-point field the source is one or more spiritual beings.** Once started, **these waves go on forever to the ends of the universe. That explains the terrific force, seething with energy, that exists today in the zero point field which keeps the heavenly bodies in their assigned orbits.** In any medium, there was an initial displacement of one of the particles. For a slinky wave, it is usually the first coil which becomes displaced by the hand of a person. For a sound wave, it is usually the vibration of the vocal chords or a guitar string which sets the first particle of air in vibrational motion. At the location where the wave is introduced into the medium, the particles which are displaced from their equilibrium position always move in the same direction as the source of the vibration. So if you wish to create a transverse wave in a slinky, then the first coil of the slinky must be displaced in a direction perpendicular to the entire slinky. Similarly, if you wish to create a longitudinal wave in a slinky, then the first coil of the slinky must be displaced in a direction parallel to the entire slinky. **Scalar waves in the zero-point field are longitudinal waves.** Scalar waves are standing waves that look like an onion in an electron. As the wave propagates outward from the center the wave fronts form a pattern similar to an onion. Each layer of the symbolic onion would be the wave front where all the corresponding nodes of the expanding scalar wave are found at any given instant. That is, for each layer of the onion, the longitudinal waves making up that particular layer have their troughs and crests in phase at that particular instant. This onion phase image, then, is the best picture of a scalar, standing wave that anyone will ever give you. A scalar wave, of necessity, must always be a standing wave. Standing waves are not absorbed or exchanged by the surroundings, while the rest are. **Exchanges are always resonant - like two piano strings of the same tone. That is about all there is to exchanges and non-exchanges.** This is the secret to the construction of our universe where all particles were/are created as scalar, standing waves. Zero-point waves of every conceivable frequency fill the vacuum of space. Every so often just like piano keys, at certain points in this frequency spectrum, scalar standing waves are allowed to develop and remain and are NOT absorbed by the surrounding wave cacophony. The Universe is very simple because ALL the rules are contained in Principle I (the Scalar Wave equation) and Principle II (medium density), and of course 'space' itself, the viewpoint of dimension. These rules create the Atomic Table and that leads to ALL the millions of different metals/atoms/molecules/compounds. We can't ever calculate them all, but the rules are so simple that we understand how it works.

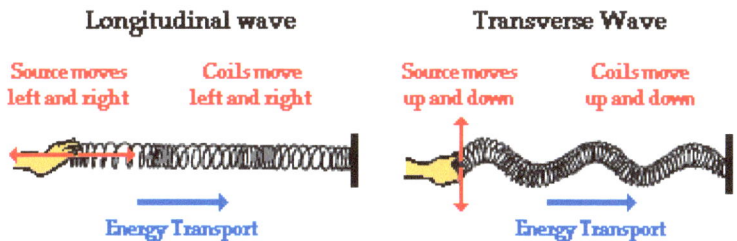

The subsequent direction of motion of individual particles of a medium is the same as the direction of vibration of the source of the disturbance.

Another way to categorize waves is on the basis of the ability (or inability) to transmit energy through a vacuum (i.e., empty space). Categorizing waves on this basis leads to three notable categories: electromagnetic waves, mechanical waves, and zero-point waves.

An **electromagnetic wave** is a wave which is capable of transmitting its energy through a vacuum (i.e., empty space). Electromagnetic waves are produced by the vibration of electrons within atoms on the Sun's surface. These waves subsequently travel through the vacuum of outer space, subsequently reaching Earth. Were it not for the ability of electromagnetic waves to travel to Earth, there would undoubtedly be no life on Earth. All light waves are examples of electromagnetic waves.

A **mechanical wave** is a wave which is not capable of transmitting its energy through a vacuum. Mechanical waves require a medium in order to transport their energy from one location to another. A sound wave is an example of a mechanical wave. Sound waves are incapable of traveling through a vacuum. Slinky waves, water waves, and telephone cord waves are other examples of mechanical waves; each requires some medium in order to exist. A slinky wave requires the coils of the slinky; a water wave requires water; a stadium wave requires fans in a stadium; and a telephone chord wave requires a telephone chord.

A zero-point wave is a product of thought and the disturbance travels on a medium called postulated (imagined) space. **Affinity** is the sympathetic vibration of two zero-point waves of thought. It is also the cohesion or adhesion of wave structures of matter. **Reality** is the standing wave of two zero-point thought waves. Everyone agrees that a solid lead brick is real; it is composed of many high frequency standing waves. Everyone also agrees that when one plucks the "A" string of a guitar, that string is the standing wave singing and vibrating. The other strings aren't in motion, they are not real. **Communication** is the energy exchange between two or more waves, zero-point or physical. Communication between beings is never passive. It can only exist where source and effect reach out to exchange energy. For example, if one person isn't actively speaking and another isn't actively interpreting the sonic energy impinging on her ear, there is no communication. Also, in radio, if a transmitter isn't actively emitting energy and a receiver isn't actively receiving, processing and converting the energy, no communication will occur. Even a spirit or cosmic being would see, hear and feel nothing if it weren't reaching to receive the energy exchange from source points.

Cosmic beings are also the life force, life energy, divine energy, élan vital, or by any other name, the energy peculiar to life, which acts upon material in the Physical Universe and animates it, mobilizes it, and changes it. Cosmic beings have no dimension or size, they are omnipresent. They are bigger than all universes combined. They are unbounded, like God, and are the source of all universes. They are you and me.

The ancient mystics knew this truth but until the development of nuclear physics its applications to life were limited. The Upanishads from the 6th Century B.C.E. contain the essential truth that "The Self is one. Unmoving, it moves faster than the mind. The senses lag, but Self runs ahead. Unmoving, it outruns pursuit. Out of Self comes the breath that is the life of all things. Unmoving, it moves things, is far away, yet near; within all, outside all. Of a certainty the man who can see all creatures in himself, himself in all creatures, knows no sorrow. How can a wise man, knowing the unity of life, seeing all creatures in himself, be deluded or sorrowful? The Self is everywhere, without a body, without a shape, whole, pure, wise, all knowing, far shining, self-depending, all transcending; in the eternal procession assigning to every period its proper duty." That is what the Upanishads say.

All zero-point waves travel at superluminal speed. They are practically instantaneous cause and effect because they have no time. Time is produced in the space resonance by slowing things down by millions of orders of magnitude. When a matter wave is produced, say an electron, everything slows down to a crawl and gravity, magnetism, charge and inertia are produced. A cosmic being with a viewpoint on Arcturus can contact an earthbound being instantly. There is no time in the thought universe. Similarly, all the atoms as wave structures in space are instantly in communication with one another through the zero-point field. The zero-point field is the medium of space resonances. All the properties of the Physical Universe including relativity, quantum mechanics, electric charge, speed of light, mass, inertia, gravity, magnetism and energy exchange are derived from the zero-point field. Each human is a spirit motivating an animal body. If we could be rehabilitated we would be cause over matter, energy, space, time and life itself. Right now most humans are unaware of their spiritual nature and their fantastic potential. The electric charge in their orbs makes them forget who they are and what they have done in the past. They won't be able to think clearly until the charge is removed.

For example, I had an upset with a fellow worker who conned me out of $2,000. I was so angry that I couldn't sleep and my attention was compulsively pinned on the thief. So I pictured the thief in my mind facing himself and the electric sparks and lightning started to fly between the two images. After a couple of minutes the lightning died down and fizzled out. I was back in control of my mind. I lost the compulsion, could think clearly again and could sleep normally.

24) HOW ARE ZERO-POINT WAVES DIFFERENT FROM PHYSICAL WAVES?

Space is common to both the thought universe and the Physical Universe. To be certain of that close your eyes and look at your house. Do you see it there? That is a house created in the thought universe and it has space. It has the space you created for it in your own universe. But it doesn't have mass, you can lift it with the little finger of you imaginary hand. It has no energy so it won't burn up and it has no time so it won't last. The second that you look away and put your attention on something else, it will disappear.

Space is only a viewpoint of dimension. Both universes have viewpoints of dimension, but they aren't the same space. An outpost in the thought universe of the federation of planets to which we belong may be situated on Venus, an uninhabitable planet per earth standards, and not be visible to us. That is because it is created in its own space. It is similar to the situation where you are sitting in your hotrod dreaming of making out with Mabel. When you open your eyes Mabel is gone because she wasn't in the seat of your hotrod in the Physical Universe. She was in the seat of your hotrod which you constructed in your own universe on top of the Physical Universe. A universe is a whole system of created things and each cosmic being creates its own universe. If a buddy borrowed your car and sideswiped it against a bridge railing, he could arrange to drive it past you smiling and waving while you see an undamaged vehicle. You saw the undamaged side of the vehicle and dubbed in the damaged, unseen side, as perfect. When you finally see the damage, it will be part of your universe.

This poem by Valerie Cox illustrates the point of contrary universes beautifully. Notice that three universes are involved here. They are your universe, the other guy's universe and the Physical Universe. All three universes can be in disagreement.

The Cookie Thief
A woman was waiting at an airport one night,
With several long hours before her flight.
She hunted for a book in an airport shop.
Bought a bag of cookies and found a place to drop.

She was engrossed in her book but happened to see,
That the man sitting beside her, as bold as could be.
Grabbed a cookie or two from the bag in between,
Which she tried to ignore to avoid a scene.

So she munched the cookies and watched the clock,
As the gutsy cookie thief diminished her stock.
She was getting more irritated as the minutes ticked by,
Thinking, "If I wasn't so nice, I would blacken his eye."

With each cookie she took, he took one too,
When only one was left, she wondered what he would do.
With a smile on his face, and a nervous laugh,
He took the last cookie and broke it in half.

He offered her half, as he ate the other,
She snatched it from him and thought... oooh, brother.
This guy has some nerve and he's also rude,
Why he didn't even show any gratitude!

She had never known when she had been so galled,
And sighed with relief when her flight was called.
She gathered her belongings and headed to the gate,
Refusing to look back at the thieving ingrate.

She boarded the plane, and sank in her seat,
Then she sought her book, which was almost complete.
As she reached in her baggage, she gasped with surprise,
There was her bag of cookies, in front of her eyes.

If mine are here, she moaned in despair,
The others were his, and he tried to share.
Too late to apologize, she realized with grief,
That she was the rude one, the ingrate, the thief.

COSMIC MECHANICS

How many times in our lives have we absolutely known that something was a certain way, only to discover later that what we believed to be true ... was not? There was a difference in reality between our own universe and the Physical Universe. This is what con artists use to deprive us of the fruits of our labors. This is what shadow governments use to deprive us of our civil rights, gold, oil, land, social security and retirement benefits. When alien technology crashes in Roswell New Mexico, the shadow government invents a weather balloon cover story so it can use the technology for its own purposes. Also, the reptilian "custodians" don't want you looking at alien bodies and technology because that will give you clues to the oppression you are experiencing. They'd rather have you watch TV to be entertained and hypnotized. When the stock market is about to crash, it is important to publish reassuring news about its strength and stability while the banks and fund managers are selling their portfolios as quietly as possible. When Martha Stewart was informed by her accountant that her stock was about to plummet, she sold out before outsiders received the news. When the CEO of Enron, Jeffrey Skilling, ran out of accounting schemes to mask the huge debt and cash flow problems that he created with pervasive fraud and conspiracy, he sold his stock before the company went bankrupt. He sentenced thousands to a life of poverty and the loss of their retirement benefits. After 9/11, President George W. Bush

took us to war in Iraq based on the false assertion that Saddam Hussein was building weapons of mass destruction. The destroyers of the World Trade Towers have been allowed to escape judgment for their crimes while President Bush uses it as a justification to increase his wartime powers. Aircraft striking the towers did not bring them down; demolition charges already in place did the job. History is replete with these con jobs. In a purely artificial chess game Roosevelt sacrificed over 2400 American seamen's lives by over-looking the obvious facts of an attack by Japan on Pearl Harbor. By this ruse, Roosevelt was able to control both the political and economic systems of the United States. These few examples of the pervasive fraud and corruption in our society should teach you that men ruled by their orbs in secret societies cannot be trusted.

So what can you do to protect yourself? You can improve your awareness! You can look around you and see that big pharma and psychiatry are in bed with the government and the reptilian "custodians." The "custodians" can only deceive the people if they are dumbed down with TV, drugs, incarcerated in mental hospitals, electro-shocked, given news with a political spin, educated to be illiterate, given an unfair justice system, given corrupt congressmen and senators, given manipulated supreme court judges, district attorney's, World Health Organization officials and Surgeon Generals and deliberate false teachings in the humanities and sciences. On this planet you can expect to be conned in all these ways and more by oppressive beings. So be on the lookout for it. The only way out is to confront these atrocities and rehabilitate our world. This last paragraph is the key to the salvation of planet earth. Find like-minded people of good will and work with them to improve conditions.

Now that you understand the relationships between the three universes, yours, the other guys and the physical, let's get back to energy. Energy is common to all universes also, because waves move in all universes. But thought energy is different than physical energy. There is no chemical, inertial, electromagnetic energy or gravity in the zero-point thought universe. You cannot bat a ball in the thought universe. You can get a mental picture of the event and see the ball land in left field, but no inertial, ballistic or aerodynamic forces come to play. The creation of mental image pictures of yourself flying across the Physical Universe and expanding in size to millions of orders of magnitude greater than the Physical Universe is done instantly in the thought universe. In the thought universe time is only have or have not, and things are simply thought forms, vibrations of the stillness. When a being's attention scans an interference pattern, such as in it's orb, an object appears, but there is no gravity or magnetism. There is electrical charge. It is like in the dream state. Things come and go, things move, things exist because you think they exist. They are just thought forms and don't have mass, gravity or inertia. Things happen instantly. It doesn't take time for eggs to cook or for you to build a house. The images that exist in the thought universe are created by the beings who live there. Those images are not controlled by the law of conservation of energy as physical waves are. Those images are postulated (imagined) images. Beings create the wave forms and say that they exist, so they do. And the motion in the thought universe does not run down over time and die out as it does in the Physical Universe. It is similar to a super-cooled state where there is no resistance to the flow of charge or energy, and energy and information flow at superluminal velocities through the stillness of space. Whereas in the Physical Universe the maxim velocity of wave structures is c, the speed of light, in the zero-point field, the minimum velocity is c, the speed of light. The speed of light divides the dualism of the thought universe and the Physical Universe. In the zero-point field energy and thought are a tangible dualism the same as mass and waves are a tangible dualism in the Physical Universe. But the source of energy in the zero-point field is intangible as is energy in the Physical Universe. In other words, humans find waves and mass tangible but energy intangible. Whereas spirits find thoughts and energy tangible, but they find the source of energy and thought is intangible. The sources are matter, energy, space and time creation units; us. Things in the spirit world are thought projections and their motion is energy which spirits can see and feel with over 60 different perceptics.

In the Physical Universe we can calculate energy, predict it, draw conclusions from it, use it to heat and cool our homes, propel engines, amplify sound and transmit radio and television signals. But until now we did not really understand the true nature and source of energy at large. It was intangible. Now we know that energy is an entity actually belonging to and originating from the thought, or zero-point universe. That's why permanent magnets will hold your notes to the refrigerator door forever. Its energy comes from the thought universe where if flows ceaselessly. Energy is a key element in the thought universe and a driving force in the Physical Universe.

Another wonderful distinction to be made between the two universes is that, although the Physical Universe is made up of space resonances which have physical properties foreign to the thought universe, each space resonance is constructed of zero-point waves which come from the thought universe. Thought is a static containing an image of motion. **The cosmic being perturbs the stillness causing scalar (longitudinal) waves which carry the energy and information.** So each space resonance is built from zero-point energy flows that have their source in thought. Energy is the real substance behind the appearance of matter and forms. The source of energy is the beings in the thought and physical universes. Thought contains images of motion; pictures. Thought has unlimited capabilities but it has no wavelength, no space and no time. It is static, infinitely still. It is impinged upon the Physical Universe which has space, energy, time and matter (space resonances). Thought is not motion in space and time. Thought is a perfect stillness containing an image of motion. Thought is not part of the Physical Universe, but it creates the Physical Universe. It controls physical energy but has no wavelength itself. It uses space resonances, but it has no mass itself. It exists in space but has no position. Only the viewpoint of a cosmic being has an assumed position. It records time but is not subject to time. Thought causes everything in an organism both structural and functional. This is why energy healing works so well and why an organism without thought is already dead.

Example of person applying this quantum technology

A person suffering from a bunion on their right foot goes to a doctor who applies 21st century 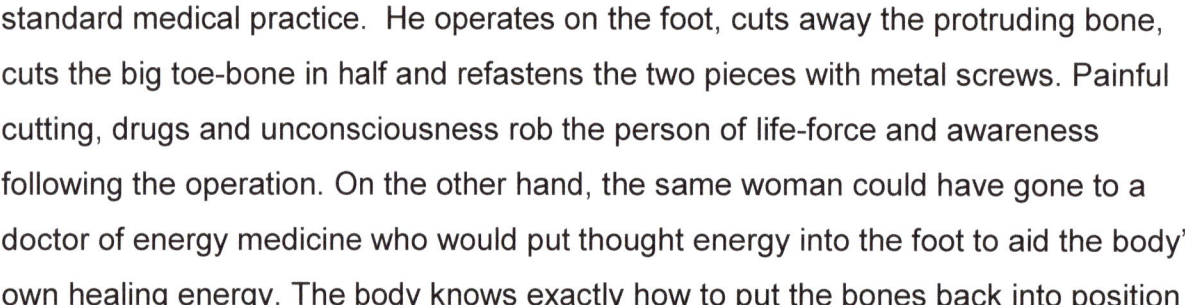 standard medical practice. He operates on the foot, cuts away the protruding bone, cuts the big toe-bone in half and refastens the two pieces with metal screws. Painful cutting, drugs and unconsciousness rob the person of life-force and awareness following the operation. On the other hand, the same woman could have gone to a doctor of energy medicine who would put thought energy into the foot to aid the body's own healing energy. The body knows exactly how to put the bones back into position and heal the tendons to return full use of the foot and cancel the pain. All that was needed was a little extra energy from another person's life-force.

The wave structure of matter contains lots of motion; subatomic motion. More complex atoms and molecules contain atomic motion, Brownian motion and flows of all kinds. Thought records these motions and times. Similarly, each of the space resonances in our orbs is associated with pictures of the events that caused us to place that mass there. Similarly, when one of the 600 million cells that die in our bodies every day is replaced, the time track of the preceding cells in that location is maintained and the new cell knows what to do based on past efforts and counter-efforts. Thought causes everything in an organism, both structural and functional. So it should come as no surprise that water has a memory, that energy healers can put beneficial frequencies in crystals and that humans have a visual aura and the power to heal. Thought is a static universe of unlimited capabilities. It impinges on the Physical Universe by means of zero-point waves producing space resonances. Its purpose is to conquer the Physical Universe to obtain ultimate survival. The best survival machine that I have ever heard of is the reptilian life form. It is superhuman, fast, strong, exceedingly intelligent, efficient, and exceptionally perceptive. Its lifespan probably exceeds 3,000 years. Lucifer of the Old Testament Bible was a serpent.

COSMIC MECHANICS

When you pluck a guitar string, it makes a nice tone which immediately loses volume until the string is quiet again. If you plucked a string of a thought guitar it would continue to sing at its original volume forever unless you expected it to die out as a physical string would do. That is its super-cooled quality. It does whatever you tell it to do, absolutely. It will sing forever just like the magnet will pull or push forever; and gravity will pull for ever. So the energies in the two universes act a little differently. There are images of motion in the thought universe and in the Physical

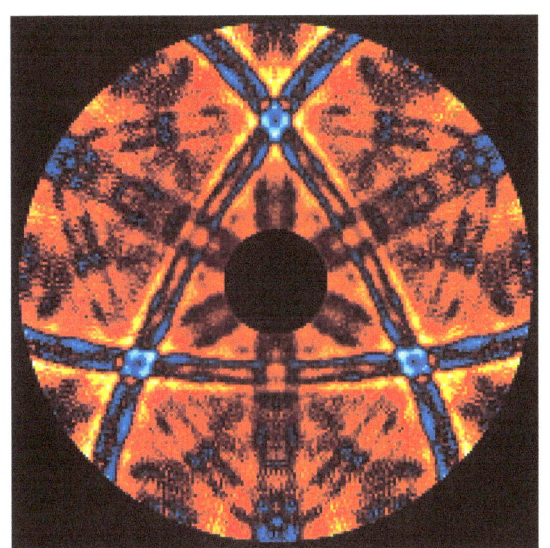

Universe energy derived from the imposition of space between things causes motion. **Affinity, Reality and Communication result in understanding in the thought universe. But Affinity, Reality and Communication result in creation in the physical universe.** The atoms in the walls that support the ceiling in your room are violently pulling together to maintain their form. They will never stop pulling; the same way that magnetism, gravity and inertia are eternal. They are postulated forces from the spiritual universe, the universe of thought.

Published in: Physical Review Letters 96, 035502 (2006)
Conventional x-ray crystallographic methods, which have been used for almost 100 years, detect x-rays diffracted by the crystal planes. In a recent paper it was demonstrated that the atomic structure can be imaged directly from real-space projections sensed by absorbing atoms inside a crystal. The use of white x-rays allowed neglecting the diffraction effects and treating the x-ray beam as a searchlight which directly produces x-ray projections of the main atomic planes in the crystal. The recorded x-ray patterns were processed in a way similar to the tomography technique which is used e.g. for 3D visualization of macroscopic objects in medical imaging.

Individual atoms are far too small to be seen in any light microscope. The inherent limitation on such a microscope is the wavelength of light, which determines the minimum size of an image. In comparison, the distance between atoms in a solid is less than 1,000th of the wavelength of light.

Notice the crystalline arrangement of the atomic structure in these pictures. Almost all non-living solid matter is crystalline but wave structures in a gas or liquid state are not. When the atoms are very far apart from each

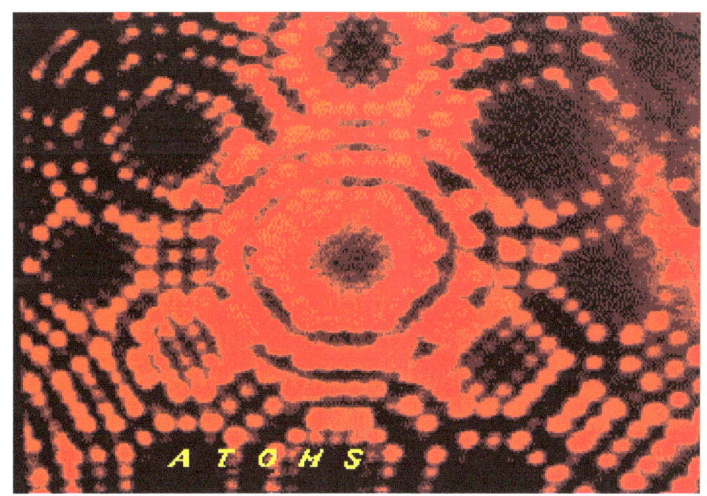

other and very disorganized – vibrating every which way – we say that the substance is a gas, as air is. When the atoms are closer together (higher density) and are attracted to each other more they make a liquid, as water is. And when the atoms are so close together that they organize themselves into rows and columns they make a solid, as rock is. This close association of atoms makes the solid rigid and crystalline. Below is a picture of the atomic structure in a metal.

25) THE ANATOMY OF A WAVE

A transverse wave is a wave in which the particles of the medium are displaced in a direction perpendicular to the direction of energy transport. A transverse wave can be created in a rope if the rope is stretched out horizontally and the end is vibrated back-and-forth in a vertical direction. If a snapshot of such a transverse wave could be taken so as to *freeze* the shape of the rope in time, then it would look like the following diagram.

COSMIC MECHANICS

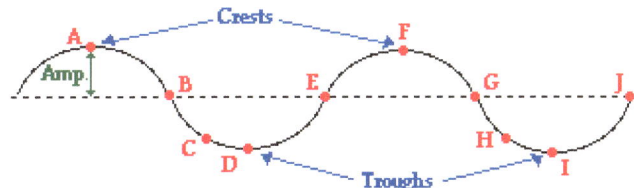

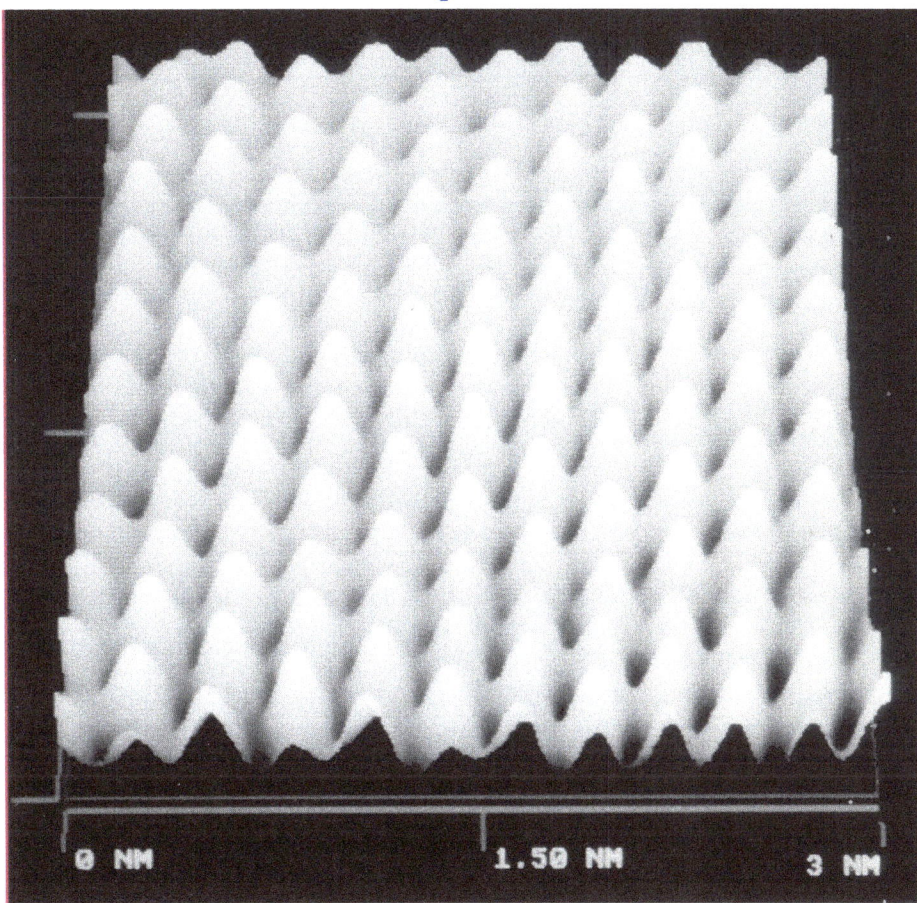

This micrograph, which represents the surface of a gold specimen, was taken with a sophisticated atomic force microscope (AFM). Individual atoms for this (111) crystallographic surface plane are resolved. Also note the dimensional scale (in the nanometer range) below the micrograph. (Image courtesy of Dr. Michael Green, TopoMetrix Corporation.)

The dashed line drawn through the center of the diagram represents the equilibrium or rest position of the string. This is the position that the string would assume if there were no disturbance moving through it. Once a disturbance is introduced into the string, the particles of the string begin to vibrate upwards and downwards. At any given moment in time, a particle on the medium could be above or below the rest position. Points A and F on the diagram represent the crests of this wave. The **crest** of a wave is the point on the medium which exhibits the maximum amount of positive or upwards displacement from the rest position. Points D and I on the diagram represent the troughs of this wave. The **trough** of a wave is the point on the medium which exhibits the maximum amount of negative or downwards displacement from the rest position.

The wave shown above can be described by a variety of properties. One such property is amplitude. The **amplitude** of a wave refers to the maximum amount of displacement of a particle on the medium from its rest position. In a sense, the amplitude is the distance *from rest to crest*. Similarly, the amplitude can be measured from the rest position to the trough position. In the diagram above, the amplitude could be measured as the distance of a line segment which is perpendicular to the rest position and extends vertically upward from the rest position to point A.

The wavelength is another property of a wave which is portrayed in the diagram above. The **wavelength** of a wave is simply the length of one complete wave cycle. If you were to trace your finger across the wave in the diagram above, you would notice that your finger repeats its path. A wave has a repeating pattern. And the length of one such repetition (known as a *wave cycle*) is the wavelength. The wavelength can be measured as the distance from crest to crest or from trough to trough. In fact, the wavelength of a wave can be measured as the distance from a point on a wave to the corresponding point on the next cycle of the wave. In the diagram above, the wavelength is the horizontal distance from A to F, or the horizontal distance from B to G, or the horizontal distance from E to J, or the horizontal distance from D to I, or the horizontal distance from C to H. Any one of these distance measurements would suffice in determining the wavelength of this wave.

A longitudinal wave is a wave in which the particles of the medium are displaced in a direction parallel to the direction of energy transport. A longitudinal wave can be created in a slinky if the slinky is stretched out horizontally and the end coil is vibrated back-and-forth in a horizontal direction. If a snapshot of such a longitudinal wave could be taken so as to *freeze* the shape of the slinky in time, then it would look like the following diagram.

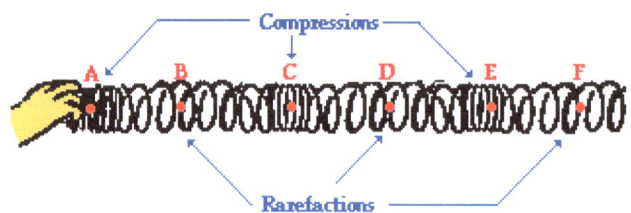

Because the coils of the slinky are vibrating longitudinally, there are regions where they become pressed together and other regions where they are spread apart. A region where the coils are pressed together in a small amount of space is known as a compression. A **compression** is a point on a medium through which a longitudinal wave is traveling which has the maximum density. A region where the coils are spread apart, thus maximizing the distance between coils, is known as a rarefaction. A **rarefaction** is a point on a medium through which a longitudinal wave is traveling which has the minimum density. Points A, C and E on the diagram above represent compressions and points B, D, and F represent rarefactions. While a transverse wave has an alternating pattern of crests and troughs, a longitudinal wave has an alternating pattern of compressions and rarefactions.

As discussed above, the wavelength of a wave is the length of one complete cycle of a wave. For a transverse wave, the wavelength is determined by measuring from crest to crest. A longitudinal wave does not have crest; so how can its wavelength be determined? The wavelength can always be determined by measuring the distance between any two corresponding points on adjacent waves. In the case of a longitudinal wave, a wavelength measurement is made by measuring the distance from a compression to the next compression or from a rarefaction to the next rarefaction. On the diagram above, the distance from point A to point C or from point B to point D would be representative of the wavelength.

FREQUENCY AND PERIOD OF A WAVE

Suppose that a hand holding the first coil of a slinky is moved back-and-forth two complete cycles in one second. The rate of the hand's motion would be 2 cycles/second. The first coil, being attached to the hand, in turn would vibrate at a rate of 2 cycles/second. The second coil, being attached to the first coil, would vibrate at a rate of 2 cycles/second. In fact, every coil of the slinky would vibrate at this rate of 2 cycles/second. This rate of 2 cycles/second is referred to as the frequency of the wave. The **frequency** of a wave refers to how often the particles of the medium vibrate when a wave passes through the medium. Frequency is a part of our common, everyday language. For example, it is not uncommon to hear a question like "How *frequently* do you mow the lawn during the summer months?" Of course the question is an inquiry about *how often* the lawn is mowed and the answer is usually given in the form of "1 time per week." In mathematical terms, the frequency is the number of complete vibrational cycles of a medium per a given amount of time. Given this definition, it is reasonable that the quantity *frequency* would have units of cycles/second, waves/second, vibrations/second, or something/second. Another unit for frequency is the Hertz (abbreviated Hz) where 1 Hz is equivalent to 1 cycle/second. If a coil of slinky makes 2 vibrational cycles in one second, then the frequency is 2 Hz. If a coil of slinky makes 3 vibrational cycles in one second, then the frequency is 3 Hz. And if a coil makes 8 vibrational cycles in 4 seconds, then the frequency is 2 Hz (8 cycles/4 s = 2 cycles/s).

The quantity frequency is often confused with the quantity period. Period refers to the time which it takes to do something. When an event occurs repeatedly, then we say that the event is **periodic** and refer to the time for the event to repeat itself as the period. The **period** of a wave is the time for a particle on a medium to make one complete vibrational cycle. Period, being a time, is measured in units of time such as seconds, hours, days or years. The period of orbit for the Earth around the Sun is approximately 365 days; it takes 365 days for the Earth to complete a cycle. The period of a typical class at a high school might be 55 minutes; every 55 minutes a class cycle begins (50 minutes for class and 5 minutes for passing time means that a class begins every 55 minutes). The period for the minute hand on a clock is 60 seconds; it takes the minute hand 60 seconds to complete one cycle around the clock.

Frequency and period are distinctly different, yet related, quantities. Frequency refers to how often something happens; period refers to the time it takes something to happen. Frequency is a rate quantity; period is a time quantity. Frequency is the cycles/second; period is the seconds/cycle. As an example of the distinction and the relatedness of frequency and period, consider a woodpecker that drums upon a tree at a periodic rate. If the woodpecker drums upon a tree 2 times in one second, then the frequency is 2 Hz; each drum must endure for one-half a second, so the period is 0.5 s. If the woodpecker drums upon a tree 4 times in one second, then the frequency is 4 Hz; each drum must endure for one-fourth a second, so the period is 0.25 s. If the woodpecker drums upon a tree 5 times in one second, then the frequency is 5 Hz; each drum must endure for one-fifth a second, so the period is 0.2 s. Do you observe the relationship? Mathematically, the period is the reciprocal of the frequency and vice versa. In equation form, this is expressed as follows.

$$\text{period} = \frac{1}{\text{frequency}} \qquad \text{frequency} = \frac{1}{\text{period}}$$

Since the symbol **f** is used for frequency and the symbol **T** is used for period, these equations are also expressed as:

$$T = \frac{1}{f} \qquad f = \frac{1}{T}$$

The quantity frequency is also confused with the quantity speed. The speed of an object refers to how fast an object is moving and is usually expressed as the distance traveled per time of travel. For a wave, the speed is the distance traveled by a given point on the wave (such as a crest) in a given period of time. So while wave frequency refers to the number of cycles occurring per second, wave speed refers to the meters traveled per second. A wave can vibrate back and forth very frequently, yet have a small speed; and a wave can vibrate back and forth with a low frequency, yet have a high speed. Frequency and speed are distinctly different quantities.

ENERGY TRANSPORT AND THE AMPLITUTDE OF A WAVE

As mentioned earlier, a wave is an energy transport phenomenon which transports energy along a medium without transporting matter. A pulse or a wave is introduced into a slinky when a person holds the first coil and gives it a back-and-forth motion. This creates a disturbance within the medium; this disturbance subsequently travels from coil to coil, transporting energy as it moves. The energy is imparted to the medium by the person as he/she does work upon the first coil to give it kinetic energy. This energy is transferred from coil to coil until it arrives at the end of the slinky. If you were holding the opposite end of the slinky, then you would feel the energy as it reaches your end. In fact, a high energy pulse would likely do some rather noticeable work upon your hand upon reaching the end of the medium; the last coil of the medium would displace your hand in the same direction of motion of the coil. For the same reasons, a high energy ocean wave does considerable damage to the piers along the shoreline when it crashes upon it.

The amount of energy carried by a wave is related to the amplitude and the frequency of the wave. A high energy wave is characterized by high amplitude; a low energy wave is characterized by low amplitude. As discussed earlier, the amplitude of a wave refers to the maximum amount of displacement of a particle on the medium from its rest position. The logic underlying the energy-amplitude relationship is as follows: If a slinky is stretched out in a horizontal direction and a transverse pulse is introduced into the slinky, the first coil is given an initial amount of displacement. The displacement is due to the force applied by the person upon the coil to displace it a given amount from rest. The more energy that the person puts into the pulse the more work which he/she will do upon the first coil. Also notice that if the person vibrates the slinky faster, more energy is put into the higher frequency waves and the person will tire quickly due to the energy being exchanged with the slinky. The more work which is done upon the first coil, the more displacement is given to it. The more displacement which is given to the first coil the more amplitude which it will have. So in the end, the amplitude of a transverse pulse is related to the energy which that pulse transports through the medium. Putting a lot of energy into a transverse pulse will affect the amplitude and wavelength but not the speed of the pulse.

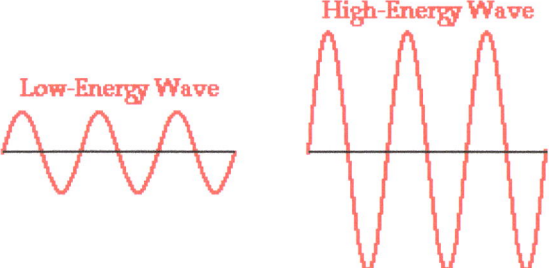

The amplitude of a wave is related to the energy which it transports.

Consider two identical slinkies into which a pulse is introduced. If the same amount of energy is introduced into each slinky, then each pulse will have the same amplitude. But what if the slinkies are different? What if one is made of zinc and the other is made of copper? Will the amplitudes now be the same or different? If a pulse is introduced into two different slinkies by imparting the same amount of energy, then the amplitudes of the pulses will not necessarily be the same. In a situation such as this, the actual amplitude assumed by the pulse is dependent upon two types of factors: an inertial factor and an elastic factor. Two different materials have different mass densities. The imparting of energy to the first coil of a slinky is done by the application of a force to this coil. More massive slinkies have a greater inertia and thus tend to resist the force; this increased resistance by the greater mass tends to cause a reduction in the amplitude of the pulse. Different materials also have differing degrees of *springiness* or elasticity. A more elastic medium will tend to offer less resistance to the force and allow a greater amplitude pulse to travel through it; being less rigid (and therefore more elastic), the same force causes a greater amplitude.

The energy transported by a wave is directly proportional to the square of the amplitude of the wave. This energy-amplitude relationship is sometimes expressed in the following manner.

This means that a doubling of the amplitude of a wave is indicative of a quadrupling of the energy transported by the wave. A tripling of the amplitude of a wave is indicative of a nine-fold increase in the amount of energy transported by the wave. And a quadrupling of the amplitude of a wave is indicative of a 16-fold increase in the amount of energy transported by the wave.

26) THE SPEED OF A WAVE

A wave is a disturbance which moves along a medium from one end to the other. If one watches an ocean wave moving along the medium (the ocean water), one can observe that the crest of the wave is moving from one location to another over a given interval of time. The crest is observed to cover distance. The speed of an object refers to how fast an object is moving and is usually expressed as the distance traveled per time of travel. In the case of a wave, the speed is the distance traveled by a given point on the wave (such as a crest) in a given interval of time. In equation form,

$$\text{speed} = \frac{\text{distance}}{\text{time}}$$

If the crest of an ocean wave moves a distance of 20 meters in 5 seconds, then the speed of the ocean wave is 4 m/s. On the other hand, if the crest of an ocean wave moves a distance of 25 meters in 5 seconds (the same amount of time), then the speed of this ocean wave is 5 m/s. The faster wave travels a greater distance in the same amount of time.

Sometimes a wave encounters the end of a medium and the presence of a different medium. For example, a wave introduced by a person into one end of a slinky will travel through the slinky and eventually reach the end of the slinky and the presence of the hand of a second person. One behavior which waves undergo at the end of a medium is reflection. The wave will reflect or bounce off the person's hand. When a wave undergoes reflection, it reverses direction 180^0 and remains within the medium. In the case of a slinky wave, the disturbance can be seen traveling into the point where the slinky is tied down and then turning back toward the original end. A slinky wave which travels to the end of a slinky and back has *doubled its distance*. That is, by reflecting back to the original location, the wave has traveled a distance which is equal to twice the length of the slinky. When subspace waves flow into an electron they spin 720^0 and then flow back out. The center of the space resonance is where the mass, inertia and charge are detected.

Reflection phenomena are commonly observed with sound waves. When you *let out a holler* within a canyon, you often hear the echo of the holler. The sound wave travels through the medium (air in this case), reflects off the canyon wall and returns to its origin (you); the result is that you hear the echo (the reflected sound wave) of your holler. A classic physics problem goes like this:

If an echo is heard one second after the holler and reflects off canyon walls which are a distance of 170 meters away, then what is the speed of the wave?

In this instance, the sound wave travels 340 meters in 1 second, so the speed of the wave is 340 m/s. Remember, when there is a reflection, the wave *doubles its distance*. In other words, the distance traveled by the sound wave in 1 second is equivalent to the 170 meters down to the canyon wall plus the 170 meters back from the canyon wall.

What variables affect the speed at which a wave travels through a medium? Does the frequency or wavelength of the wave affect its speed? Does the amplitude of the wave affect its speed? Or are other variables such as the mass density of the medium or the elasticity of the medium responsible for affecting the speed of the wave?

The speed of a wave is not dependent (causally affected by) the properties of the wave; rather the speed of the wave is dependent upon the properties of the medium. A wave is a disturbance moving through a medium. There are two distinct objects in this phrase - the "wave" and the "medium." The medium could be water, air, or a slinky. These media are distinguished by their properties - the material they are made of and the physical properties of that material such as the density, the temperature, the elasticity, etc. Such physical properties describe the material itself, not the wave. It should raise your affinity for the walls and furniture around you to know that the atoms of those objects are made of the stuff that you create as a spiritual being. Their physical properties are your properties as a spiritual being with the addition of time. Time makes everything in the Physical Universe run in extreme slow motion compared to thought. We are all spiritual beings. We comprise the fabric or medium of space, that infinite stillness which underlies all physical existence. Since we are infinite beings, disturbances or waves that we create travel through us at infinite speeds. You are in perfect communication with everything in your universe. You have total knowingness of everything in your own universe. Since your universe is infinite, if you have something going on in a neighboring galaxy, you know about it instantly even though you have something going on with your body on Earth at the same time.

You cannot be humanoid to be an administrator of an organization as large as a galaxy. Of necessity the Galactic Federation is primarily a thought universe organization. You wouldn't see their spaceships flying around. Humans have amnesia; they can barely remember what they had for breakfast. They cannot remember the torturous journey that wound them up in 21st century Earth. They were so irresponsible that they decided to get a body to create for them instead of doing the creating themselves. They were so stupid that they stored all their pain, unconsciousness and decisions to not-be in a little 21 gram orb that mystifies them to this day. A corporation of physical beings could not govern a galaxy. Physical beings are transient, participating only for brief periods in the physical realm. Even if we lived to 200 years we would only have a fleeting period of adult productivity. The galaxy evolves in a time frame that is fantastically long compared to a human lifespan. In order to monitor and direct the evolution of a galaxy of life, the administrators must have an active memory much greater than 200 years. The dramatic unfolding of only one specie's evolution on a planet often consumes millions of years. To direct this evolution the beings involved must be around for a long time or maintain their memory like the cells in our body do or the gray aliens do with a hive mentality. When a gray dies, as in the Roswell crash, the body perishes but the mind goes on controlling the rest of the hive.

To communicate with the outposts of the galaxy, radio and television systems cannot be used. Electromagnetic energy travels at the speed of light, 186,000 miles per second. That sounds fast, but for a star that is 16 million light years out from the galactic hub, each portion of a message would take 16 million years to cross the space and the answer back would take another sixteen million years. So obviously this system is unbearably slow. The Federation uses beings at each of its strategic outposts to communicate over vast amounts of space. There is no time in the thought universe; everything is in now, so a message sent mentally from Arcturus is received instantly on the Earth telepathically. Then the being receiving the message can act on it in the physical universe. Galaxies evolve over trillions of years and constant attention must be given to the development of programs to optimize the agenda.

That child's little orb (left) contains all the ways that beings deny themselves. It is the single source of all their problems, stress, unhappiness and self doubt. They are totally lost and when some bright individual comes along to show them the way out of the quagmire they get an urge to hang her. It looks like they don't want to be free. They hang on to the security of their prison. The truth is that the little 21 gram orb that they have been creating for 200 trillion years that serves as their viewpoint that they take with them between lives, has grown more destructively powerful than they are and now it is their public enemy number one. It is truly an insidious demon.

Here is an interesting datum that will help you understand the orb: The orb is to you as the Physical Universe is to God. As above, so below. We are playing around in God's orb. God's orb is all powerful, ours is powerless. The Physical Universe has all the force, we have none. There is a way to level the playing field, but it takes a lot of intention to achieve it.

Now back to waves. Waves are distinguished from each other by their properties - amplitude, wavelength, frequency, etc. These properties describe the wave, not the material through which the wave is moving. Wave speed depends upon the medium through which the wave is moving. Only an alteration in the properties of the medium will cause a change in the speed. In the zero-point field, where there is no time, scalar waves propagate at an infinite velocity, the speed of thought. So now you know some strange things about zero-point waves: 1) They travel at infinite velocity. 2) They don't lose their energy by dampening out. 3) Every space resonance is instantly in communication with every other space resonance in the universe by means of in and out waves.

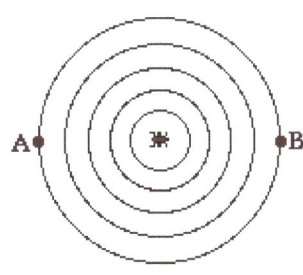

A stationary bug producing disturbances in water.

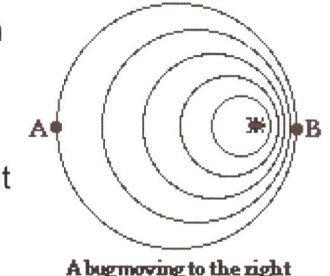

A bug moving to the right and producing disturbances.

27) THE DOPPLER EFFECT

Suppose that there is a happy bug in the center of a circular water puddle. The bug is periodically shaking its legs in order to produce disturbances that travel through the water. If these disturbances originate at a point, then they would travel outward from that point in all directions. Since each disturbance is traveling in the same medium, they would all travel in every direction at the same speed. The pattern produced by the bug's *shaking* would be a series of concentric circles as shown in the diagram at the left. These circles would strike the edges of the water puddle at the same rate. An observer at point A (the left edge of the puddle) would observe the disturbances to strike the puddle's edge at the same frequency that would be observed by an observer at point B (at the right edge of the puddle). In fact, the frequency at which disturbances reach the edge of the puddle would be the same as the frequency at which the bug produces the disturbances. If the bug produces disturbances at a frequency of 2 per second, then each observer would observe them approaching at a frequency of 2 per second.

Now suppose that our bug is moving to the right across the puddle of water and producing disturbances at the same frequency of 2 disturbances per second. Since the bug is moving towards the right, each consecutive disturbance originates from a position which is closer to observer B and farther from observer A. Subsequently, each consecutive disturbance has a shorter distance to travel before reaching observer B and thus takes less time to reach observer B. Thus, observer B observes that the frequency of arrival of the disturbances is higher than the frequency at which disturbances are produced. On the other hand, each consecutive disturbance has a further distance to travel before reaching observer A. For this reason, observer A observes a frequency of arrival which is less than the frequency at which the disturbances are produced. The net effect of the motion of the bug (the source of waves) is that the observer towards whom the bug is moving observes a frequency which is higher than 2 disturbances/second; and the observer away from whom the bug is moving observes a frequency which is less than 2 disturbances/second. This effect is known as the Doppler Effect.

The Doppler Effect is observed whenever the source of waves is moving with respect to an observer. The **Doppler Effect** can be described as the effect produced by a moving source of waves in which there is an apparent upward shift in frequency for observers towards whom the source is approaching and an apparent downward shift in frequency for observers from whom the source is receding. It is important to note that the effect does not result because of an <u>actual</u> change in the frequency of the source. Using the example above, the bug is still producing disturbances at a rate of 2 disturbances per second; it just appears to the observer whom the bug is approaching that the disturbances are being produced at a frequency greater than 2 disturbances/second. The effect is only observed because the distance between observer B and the bug is decreasing and the distance between observer A and the bug is increasing.

The Doppler effect can be observed for any type of wave - water wave, sound wave, light wave, etc. We are most familiar with the Doppler Effect because of our experiences with sound waves. Perhaps you recall an instance in which a police car or emergency vehicle was traveling towards you on the highway. As the car approached with its siren blasting, the pitch of the siren sound (a measure of the siren's frequency) was high; and then suddenly after the car passed by, the pitch of the siren sound was low. That was the Doppler Effect - an apparent shift in frequency for a sound wave produced by a moving source.

COSMIC MECHANICS
The Doppler Effect for a Moving Sound Source

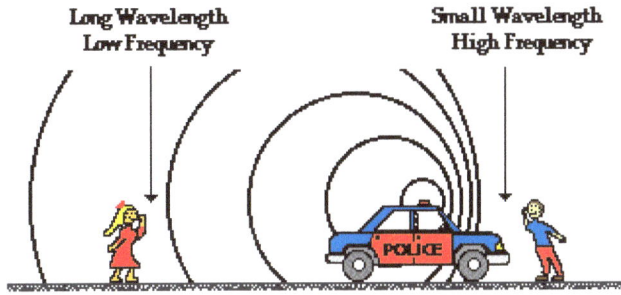

The Doppler Effect is of intense interest to astronomers who use the information about the shift in frequency of electromagnetic waves produced by moving stars in our galaxy and beyond in order to derive information about those stars and galaxies. The belief that the universe is expanding is based in part upon observations of electromagnetic waves emitted by stars in distant galaxies. Furthermore, specific information about stars within galaxies can be determined by application of the Doppler Effect. Galaxies are clusters of stars which typically rotate about some center of mass point. Electromagnetic radiation emitted by such stars in a distant galaxy would appear to be shifted downward in frequency (a "red shift") if the star is rotating in its cluster in a direction which is away from the Earth. On the other hand, there is an upward shift in frequency (a "blue shift") of such observed radiation if the star is rotating in a direction that is towards the Earth.

Another interpretation of the red shift is due to gravity. In Physics, light loses energy when it moves away from a massive body such as a star or a black hole. This effect reveals itself as a **gravitational red shift** in the frequency of the light, and is observable as a shift of spectral lines towards the longer, or "red," end of the spectrum. This is why astronomers got the idea that the universe was expanding from a "big bang." They incorrectly assumed that the red shift was due to the doppler effect rather than to a gravitational effect. Gravitational redshift is sometimes known as the Einstein effect, although that is not the only meaning applied to that term.

Light coming from a region of weaker gravity shows a gravitational blueshift.

28) TRAVELING WAVES VS. STANDING WAVES

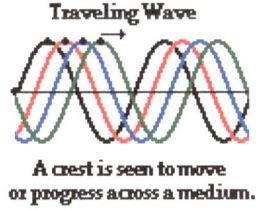

A mechanical wave is a disturbance which is created by a vibrating object and subsequently travels through a medium from one location to another, transporting energy as it moves. The mechanism by which a mechanical wave propagates itself through a medium involves particle interaction; one particle applies a push or pull on its adjacent neighbor, causing a displacement of that neighbor from the equilibrium or rest position. As a wave is observed traveling through a medium, a crest is seen moving along from particle to particle. This crest is followed by a trough which is in turn followed by the next crest. In fact, one would observe a distinct wave pattern (in the form of a sine wave) traveling through the medium. This sine wave pattern continues to move in uninterrupted fashion until it encounters another wave along the medium or until it encounters a boundary with another medium. This type of wave pattern which is seen traveling through a medium is sometimes referred to as a **traveling wave**.

Traveling waves are observed when a wave is not confined to a given space along the medium. The most commonly observed traveling wave is an ocean wave. If a wave is introduced into an elastic chord with its ends held 3 meters apart, a wave introduced on one end becomes confined in a small region. This wave will quickly reach the end of the chord, reflect and travel back in the opposite direction. Any reflected portion of the wave will then interfere with the portion of the wave incident towards the fixed end. This interference produces a new shape in the medium which seldom resembles the shape of a sine wave. Subsequently, a traveling wave (a repeating pattern which is observed to move through a medium in uninterrupted fashion) is not observed in the chord. Indeed there are traveling waves in the chord; it is just that they are not easily detectable. In such instances, rather than observing the pure shape of a sine wave pattern, a rather irregular and non-repeating pattern is produced in the chord which tends to change appearance over time. This irregular looking shape is the result of the interference of an incident sine wave pattern with a reflected sine wave pattern in a rather non-sequenced and untimely manner. Both the incident and reflected wave patterns continue their motion through the medium, meeting up with one another at different locations in different ways. For example, the middle of the chord might experience a crest meeting a *half crest*; then moments later, a crest meeting a *quarter trough*; then moments later, a *three-quarters crest* meeting a *one-fifth trough*, etc. This interference leads to a very irregular and non-repeating motion of the medium. The appearance of an actual wave pattern is difficult to detect amidst the irregular motions of the individual particles.

It is possible however to have a wave confined to a given space in a medium and still produce a regular wave pattern which is readily discernible amidst the motion of the medium. For instance, if an elastic rope is held end to end and vibrated at just the right frequency, a wave pattern would be produced which assumes the shape of a sine wave and is seen to change over time. The wave pattern is only produced when one end of the rope is vibrated at just the right frequency. When the proper frequency is used, the interference of the incident wave and the reflected wave occur in such a manner that there are specific points along the medium which appear to be standing still. Because the observed wave pattern is characterized by points which appear to be standing still, the pattern is often called a **standing wave pattern**. There are other points along the medium whose displacement changes over time, but in a regular manner. These points vibrate back and forth from a positive displacement to a negative displacement; the vibrations occur at regular time intervals such that the motion of the medium is regular and repeating - a pattern is readily observable.

The diagram at the right depicts a standing wave pattern in a medium. A snapshot of the medium over time is depicted using various colors. Note that point A on the medium moves from a positive to a negative displacement over time. The diagram only shows one-half cycle of the motion of the standing wave pattern. The motion would continue and persist, with point A returning to the same positive displacement and then continuing its back-and-forth vibration between the up to the down position. Note that point B on the medium is a point which never moves. Point B is a point of no displacement; such points are known as **nodes**. The standing wave pattern which is shown at the right is just one of many different patterns which could be produced within the rope.

29) BELL'S THEOREM

There are many aspects of modern physics which seem to violate our intuition. The classical theory of physics which was held from the time of Newton until this century provided an orderly model of a world made of objects moving around and pushing each other around in predictable ways. The mathematics could be difficult, but the basic ideas meshed with our common sense notions of how the world works. Starting at the beginning of this century, our physical theories began to include aspects which ran counter to that common sense, and yet the theories consistently made accurate predictions of experiments which could not be explained with Newtonian physics. Gradually, and despite much resistance, physicists have been forced to accept these new results.

We are all far more connected to one another than the consensus would have us believe. Our family and friends influence our well being more than we realize. Economically we see the problems on one continent may immediately impact all world markets. Globally, we have only one big ocean and all share the same water, the same air, and the same earth. Our lives and fate are inextricably linked. Even the very smallest of particles are linked in a surprisingly profound manner.

Bell's Theorem is a mathematical construct which non-mathematicians have a difficult time getting their wits around. Its implications however, affect profoundly our basic world view. Some physicists believe that it is the most important single discovery in the history of physics. One of the implications of Bell's Theorem is that, at a deep and fundamental level, the "separate parts" of the universe are connected in an intimate and immediate way. Physicists have what they call a two-particle system of zero spin. This means that the spin of each of the particles in the system cancels the other. If one of the particles in such a system has a spin up, the other particle has a spin down. If the first particle has a spin right, the second particle has a spin left. No matter how the particles are oriented, their spins are always equal and opposite. Some quantum physicists have become upset by the fact that paired electrons, traveling away from each other at the speed of light, are able to somehow respond to one another instantaneously, infinitely faster than the speed of light.

If these two particles are sent in opposite directions, no matter how far apart they may become, they are still linked. They could be thousands of light years away from one another, but if one of the particles goes through a magnetic device that changes its spin, say from up to down, the other particle, regardless of distance, will instantaneously and spontaneously change its spin from down to up. This is the concept of nonlocality. The movement of one atom in this galaxy can immediately affect one in another galaxy, as long as they began by being associated in a way called **entanglement**. Bell showed that this action at a distance is not just an idea, but an essential aspect of reality.

The converse of nonlocality is locality. It is the principle that an event which happens at one place can't instantaneously affect an event someplace else. For example: if a distant star were to suddenly blow up tomorrow, the principle of locality says that there is no way we could know about this event or be affected by it until something, *e.g.* a light beam, had time to travel from that star to Earth. Aside from being intuitive, locality seems to be necessary for relativity theory, which predicts that no signal can propagate faster than the speed of light.

In 1935, several years after quantum mechanics had been developed, Einstein, Podolsky, and Rosen published a paper which showed that under certain circumstances quantum mechanics predicted a breakdown of locality. Specifically, they showed that according to the theory, a particle could be put into a measuring device at one location and, simply by doing that, instantly influence another particle arbitrarily far away. They refused to believe that this effect, which Einstein later called "spooky action at a distance," could really happen, and thus viewed it as evidence that quantum mechanics was incomplete.

Almost thirty years later John S. Bell, an Irish physicist, proved that the results predicted by quantum mechanics could not be explained by any theory which preserved locality. In other words, if you set up an experiment like that described by Einstein, Podolsky, and Rosen, and you get the results predicted by quantum mechanics, then there is no way that locality could be true. Years later the experiments were done, and the predictions of quantum mechanics proved to be accurate. In short, one of the great pillars of science, namely the idea of locality, is dead.

This result demonstrates one of the other strange results of modern physics, which is that the act of measuring a property always changes the system you are measuring. In this case the "system" apparently includes not only the electron you are measuring, but also the other entangled one which isn't even there at the time. Physicists have been trying for over fifty years to understand these results, and there is no consensus on how to interpret them. There is clear agreement, however, that the results occur. Spooky action at a distance is part of nature. This entanglement feature of wave structures may help us to understand the creation of an orb. During periods of intense emotional distress, high energy wave structures in our own universe may entangle with wave structures in the Physical Universe to add mass and recordings to our orbs. These masses, being permanent, stay with us for trillions of years or until we realize that we created them and take responsibility for them. At which point they will vanish. The act of observation in our own universe collapses the wave function as Erwin Schroedinger would say. This act also collapses the wave function of the entangled wave structure in the Physical Universe. It no longer exists and cannot be photographed with a digital camera. It does not follow the laws of conservation of energy by converting to another form per Einstein's famous law, $E=mc^2$. It just ceases to exist and leaves no traces, period. This is the secret to erasing our orbs and removing their diabolical influence.

30) CREATION OF THE PHYSICAL UNIVERSE

The Physical Universe was created by all of us. How that was accomplished I will explain shortly. The basic motivation behind the creation of this universe was the desire to create effects. There is no other reason. We just wanted to play. We wanted a game. So we created a universe which we experience today as matter, energy, space and time. We could have created a universe based on any qualities or formulas we chose, but we chose the laws of this universe because of its interesting effects. It has been very successful, has been growing, and beings have been joining in ever since. The goal of the beings that created this universe was to produce effects upon others and to obtain good effects from others, a goal which is very similar to our own today. This explains why so many beings joined this universe. That's why this universe exists today, to produce an effect. This Physical Universe is a product of the thought universe. That is why the Physical Universe has so much thought in it. It is composed of matter, energy, space, time and thought.

31) DOES QUANTUM MECHANICS APPLY TO HUMANS?

Niels Bohr, Schrödinger, and Heisenberg demonstrated quantum effects in the laboratory with non-living subatomic particles. But what of the living? What of the DNA in cells and their biophotonic communication systems? Are they quantum systems also? The quantum physicists had discovered that our involvement with quantum particles was crucial. Subatomic wave structures existed in all possible states until disturbed by us, observed by us. At which point they would resolve into something real. Human consciousness was somehow central to turning the subatomic quantum waves into reality. It was all so counter-intuitive and strange! How could electrons be in touch with everything at once? How could an electron not be a discrete set thing until it is examined or measured? How could anything be concrete in the world if it were only agitations within agitations? Now we know that even the skin of our bodies is composed of cells built and managed by quantum energy fluxes. It is now seen that the very underpinning of our universe is one vast, heaving sea of energy and interference patterns, quantum waves. Human perception occurs because of energy exchanges between the membranes in our cells and the quantum energy sea. We resonate with the Physical Universe the same as a string sympathetically vibrates in unison with another string tuned to its frequency. Then we pick up the information in the thought universe which is our home. Thought is separate from the Physical Universe (Matter, Energy, Space and Time). Thought can operate in and with matter. Thought can consider itself integrated with matter, thought can consider itself to be matter, but in truth creative thought and perception reside in thought, not matter. So brain (matter) could never be a mind. The consciousness of the observer (thought) brings the observed object into being. We are creating our world micro-second by micro-second, twenty-four hours per day. But almost none of us realize it! What we think matters (i.e. a verb – to make matter out of thought).

This is the key to all information processing and exchange in our world. It is the key to perception and inter-cellular communication. These experiments proved the collective consciousness and the life force flowing through the universe. We are starting to see that there is a reasonable explanation for the religious truths we had believed in for centuries. But now we are finding certainties. A leap of faith is no longer required. We now know about life after death and the effectiveness of alternative medicine and prayer. We've now sort of come up with a science of religion.

32) SOLID UNIVERSES

The Physical Universe is a finite object. If you were to travel to the edge of it, you would discover that there are no more stars and galaxies beyond it. The artists drawing at the right is a remote viewer's impression of the Physical Universe seen from far away. It looks like an elongated, brown, sugar-candy with bright little lights (stars) sprinkled generously inside. A remote viewer is a person who uses scientific protocols to obtain mental images of things in other spaces and times. The thought universe is composed of images of motion and out of it comes the Physical Universe. So every being with a mind has the ability to scan the past, present and future in the thought universe for pictures he's interested in.

Two or more universes can exist right on top of one another. This is because each has a separate viewpoint of dimension. There are, actually, universes without number, but humans have their attention forcefully stuck in this Physical Universe and are not free to roam in other spaces and times. In the Physical Universe, no two objects can exist in the same space, but each of us can create a different space which won't interfere with the other created spaces. Only when many beings go into agreement with a viewpoint of dimension, such as the Physical Universe, do we get a common universe, where the space is shared or common to all.

SOLIDITY: Solidity in the Physical Universe is concerned with space. Matter is a condensation of energy that cosmic beings placed in space. The more the energy is condensed, the less space it occupies and the greater its endurance (time) becomes. Things are solid only because we say they are. A friend of mine told me of an experience that he had with a highly evolved spiritual master. The master instructed him to look at the wall but not look at the wave structures. He told my friend to concentrate on the space between the structures. To his surprise, he could see someone reading on the other side of the wall! He quickly ran into the adjoining room and saw that he had correctly observed the person reading on the other side of the wall.

There is nothing actually still or static in the Physical Universe. Each atom is found to be nothing more than particles composed of agitations within agitations. Our most powerful tunneling electron microscopes reveal that there is no actual substance to an atom. The energy simply bounces around. We have atomic motions, molecular motions and flows of all kinds. Nothing is still or at rest in the Physical Universe. Things are only solid because we consider that they are solid. The molten core of the earth, composed of heavy metals highly compressed at millions of pounds per square inch, is very solid. The air at the surface of planet Earth is compressed by only 15 pounds per square inch and isn't very solid. However, it is solid enough to hold a 500,000 pound aircraft above the surface.

We create space, simply by assuming a viewpoint and putting out points to view. This is done in the thought universe by creating ridges with our attention. These ridges are standing waves caused by two zero-point energy flows hitting each other and causing a chaotic mixture which hangs in time. Simultaneous with the creation of time, space, charge, gravity, inertia and vibrating energy is created with a spin. An example would be an electron. The action of an electron is reaching and withdrawing. There is an in wave and an out wave. There is really nothing to it. It just sits there wiggling, spinning, reaching and withdrawing. The properties of mass and charge come about due to the nonlinearity of the dense wave in the center of the structure. The structure has a vibrational frequency that other structures can sympathetically synchronize with. That makes it real. The structure tries to get away from other similarly charged electrons. That's its affinity. It continuously shares its energy with other particles at superluminal velocities carrying information. That is its communication. This reality, affinity and communication add up to existence in the Physical Universe. There is connection and interchange between the cosmic being and the electrons which is the communication that exists between the being and its creation. This is a postulated connection and interchange which is independent of Physical Universe laws such as Conservation of Energy and Inertia. These zero-point waves carry information, pictures, understandings as well as ridges, flows and dispersals. A cosmic being senses and feels with the waves which it produces. When a cosmic being produces a mental image picture of a fairy tale princess, it can see and hear from that viewpoint because the waves and ridges of the princess are in that spiritual beings universe. Everything in its own universe is part of it –because it is an infinite being without matter, energy space or time, but it can produce these. I use the pronoun "it" because spiritual beings have no gender. They are beings like you and me but they aren't trapped in a body. If you could get out of your body, you would regain your immortality. You would see that you are you and not your body. You cannot die, you are immortal. You are a spirit. You have your own universe. A universe is a whole system of created things. Just close your eyes and scan mentally through the activities of your past and you will be getting familiar with your own universe. The fact that you can look without the use of your body's eyes shows you that you have spiritual vision. You are a spiritual being. Yours is a universe very different from the Physical Universe. The more time and attention you give to your own universe, the stronger it becomes. The more time and attention you give to the Physical Universe, the stronger it becomes. So spend your time 24/7 creating things in the Physical Universe if you want to be wasted as a being. Or spend most of your time working in your own universe undoing your orb and helping others to undo theirs, and you will regain your use of force.

It is absolutely essential for us to realize that there are at least two forms of life like there are two forms of energy; physical and thought. We are composite beings. Our physical forms are inhabited by thought life-forms. Physical Life-forms are temporary creatures, they live and die. The thought aspects of these life-forms exist forever. They are the cosmic beings that existed before they became physical and will continue to live long after our physical bodies decay and are recycled in the earth. They will take their orbs with them for their next incarnation until their orbs are vanquished; which will free the being from its diabolical enslavement.

Two universes can exist right on top of each other
Close your eyes and imagine that you are sitting in the driver's seat of your car. Use your right hand to reach down to touch the radio volume control. Now open your eyes and realize that your hand is suspended in the space of the Physical Universe and that your universe is co-existing right on top of the Physical Universe. In the Physical Universe, no two wave structures can exist in the same space, but any number of universes can exist superimposed on each other at any point in the Physical Universe. Each of us has our own universe, but we interface in the common universe, the Physical Universe.

In the Physical Universe, depending on the frequency, amplitude and wave characteristics of the "particles", we get light, energy, matter, time and life! Physicists study the properties of elementary wave structures with cloud chambers, high energy linear accelerators, cyclotrons, klystrons, betatrons and tunneling electron microscopes and they classify their strange observations as virtual pi-mesons, gravitons and muons, etc., because they don't know what they are looking at. It is all an illusion. That's what makes playing with strong magnets so mystifying. **Our reality is the inevitable average of illusions in the Physical Universe. That's the same thing as the physicists saying that matter appears where the standing waves and amplitudes of the quantum wave function manifest.**

Each cosmic being (like you and me) has a viewpoint. The viewpoint can never be destroyed, but the illusion can. All humans are viewpoints too. When the viewpoints see that the creation (their bodies) can be destroyed, they assume that they can be destroyed too. Thus we fear death on this planet. Note that the particles are only an illusion, as are all particles, atoms, planets and stars in the entire world! The reality of the Physical Universe is only the inevitable average of illusions produced by the cosmic beings. It is the interplay of their quantum waves that produces the world as we know it. When you really get this, the beams and rafters of the Universe will shatter and you will be able to disagree with every part of it and regain your power. **The Physical Universe is only an illusion! But it is a powerful illusion until you break your agreement with it.** To break your agreement with it you must discover who you are and how you tricked yourself into giving all your power to it. It is difficult for humans to know who they are because they cannot see themselves, a nothingness. They only see the Physical Universe, a somethingness. So they get confused. They also get mislead by the "particle" appearance of wave structures. It is all an illusion. Then the psychiatrists, pharmaceutical companies and scientific materialism finish us off. We wind up agreeing that we are evil animals evolved from mud. We are lost souls.

The zero-point waves created by the spiritual being will disappear if the being looks away. Then when it looks back it will have to re-create them if it wants to do more with them. It is like when you dream. When you wake up, you forget about the dream. But you can recreate the dream and investigate every part of it in your mind at a later time. Notice that the Physical Universe and your own universe in your imagination are created in the same way. It is all the creation of illusions. First you get the idea in your imagination. This has no permanence or duration. Then you postulate it; that is like posting a notice in your mind that it is now true for you. Then you put out attention units to build the quantum creation. If you have mastered force you can then put this creation in the Physical Universe. This is when it first exists in matter, energy, space and time, but it isn't very solid or real at all. Then you make it durable by getting agreement on it and lying about it so that it is difficult to undo. A hypnotist will perform this operation for you by dominating your will. He'll give you the idea that everyone in the audience is naked. He'll tell you that everyone is naked and you'll go into thorough agreement with it. So you'll put out attention units to build the postulated environment and embarrass yourself. Everyone will remain naked for you until the lie is uncovered that it was all a hypnotic suggestion. Then the audience will be clothed again.

The acceleration of gravity, magnetic attraction and repulsion, and static electric attraction and repulsion do not make or consume energy. This is energy of the first class. This primary energy was put into the Physical Universe by the cosmic beings and will never run out. It is postulated energy. It is like god saying, "Let there be light." And it is forever. Your fridge magnet will work to hold your papers tightly to the door forever without ever letting go. That is because permanent magnets have electrons that continuously apply the minimum amplitude principle, pushing them into the steel refrigerator door. Gravity will forever press your body toward the center of the earth with no consumption of fuel because cosmic beings said it would be that way. It is a property that is postulated into every wave structure of the universe. Humans are worried about an energy shortage and economic collapse. But it could never happen because a cosmic being, or life-static, is a matter/energy/space/time production unit! It is impossible to ever run out of energy because the cosmic beings put it in every smallest part of this Universe, and every spiritual being like you and I can make more of it. Some enlightened people remember that spiritual beings have been used in the distant past to create energy for their slave masters.

33) GRAVITY

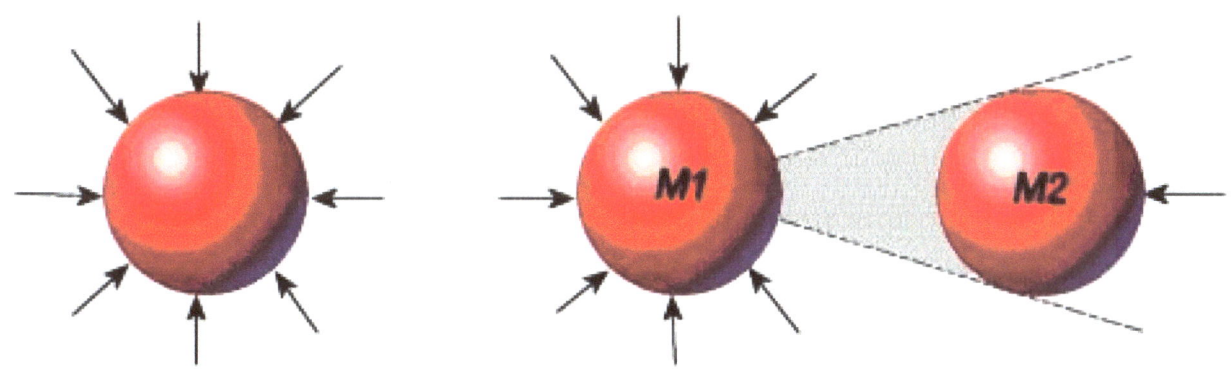

Scalar Wave Radiation Pressure Acting on Space Resonances (Masses) In Accordance With The Minimum Amplitude Principle To Produce Gravity

A simplified way to think of the effect of the Minimum Amplitude Principle as it affects gravity is shown by the above diagram. The shadowing effect of the wave structure of matter causes pressure differences known to obey the inverse square law, as in Newton's gravitational force, providing both a push from the illuminated high pressure side, and a pull from the shadowed low pressure side. The above diagram shows how mass M2 shadows the scalar quantum waves that would reach M1 in its absence. The net force vector on M1 will result in its movement towards M2. In a similar fashion M1 will cast a shadow on M2 (not shown for the sake of clarity), which will result in the motion of M2 towards M1. Similarly, inertia will be found to follow the same shadowing effect if the scalar waves from all the wave structures in the universe are combined to exert a force on the accelerating M2, per Mach's Principle.

What is gravity?

We can feel gravity acting on all bodies made up of matter. We also have two laws of gravity; Newton's and Einstein's. Newton's law simply states that every object in the universe attracts every other object with a force which for any two bodies is proportional to the mass of each and varies as the inverse square of the distance between them. This statement is expressed mathematically by the well known equation: $F = (G \, m \, m')/r^2$

Despite the success of the above law to describe the gravitational force, Newton still found himself unable to offer any adequate explanation of what and how all is happening. Also, later on, in spite of all excitement following Newton's law of universal gravitation, Einstein modified it to take into account his theory of relativity. Newton's law assumes the gravitation effect to be instantaneous, that means that a mass would 'feel' the movement of another distant mass instantaneously, without any propagation delay, and this would make gravity information go faster than the speed of light. Einstein's law thus tweaked Newton's equation to take these delays into account. Gravitational force is being applied to each and every space resonance making up matter and space itself. It can be considered as an interaction between an external force and each elementary unit which combine under different configurations to make up different kinds of matter. We can shield electric field, magnetic fields, electromagnetic fields but not gravity. Why is that? Because scalar zero-point waves penetrate everything in space and only the nonlinearities

produced in space resonances block them. (Photo above by F. Lee Aeilts)

Mass is a problem. Two flows discharging against one another produce the problem. This problem becomes a knot or nonlinearity in the smooth flow of the zero-point energy. It is like turning two fire hoses against each other. Where the flows meet in the center is a confused blob or ridge which persists while the water is flowing down the drain. One can view a ridge by turning on the tap in a kitchen sink as shown above. The water spreads in every direction until it meets the ridge of stopped flow from the previous flow which is backing up. A wave crashing on the beach is another example of a ridge. It is that ridge in the center of each electron and atomic wave structure that reacts with other ridges to produce gravity. The more ridges, the more gravity. There is increasing gravity with increasing mass. Our Moon's mass is developing a force that draws the Moon toward the mass of the Earth. At the same time, the Moon's mass is flying through the mass of all the stars, which, per Mach's principle, tends to keep it going in a straight line. In this way orbits are formed.

The property of inertia possessed by all objects in the Physical Universe is simply the resistance to being accelerated through the Zero Point Field. The larger the object, the more particles (space resonances) it contains, and the more it prevents acceleration. The corporeal stuff we call matter is an illusion. The background sea of zero-point energy opposes acceleration of the nonlinearities at the center of wave structures. Gravity works similarly. As shown in the diagram above, M2 shields M1 from a portion of the scalar Zero Point Field and vice versa. Thus the caldron of seething background energies pushes the masses together. Gravity is a pushing force, not a pulling force.

34) MECHANICS OF ENERGY FLOWS

Energy becomes matter if condensed, as in the core of the Sun, and conversely, matter becomes energy if dispersed, as in a nuclear blast. The Physical Universe is an accumulation of created energy in chaos. All the stars are pulling in mass from space which gets converted to energy in fission and fusion processes. The Physical Universe holds on to energy, won't let any of it go, but only allows it to convert between energy and matter according to Einstein's equation, $E=mc^2$. The equation has always implied that one distinct physical entity (energy) turns into another distinct physical entity (mass). But with the understanding of the wave structure of matter we see that matter is not a fundamental property of physics. The Einstein equation just quantitatively described the amount of energy necessary to create the appearance of mass. Mass is condensed energy. There aren't two; just one fundamental physical entity – energy. Any solid chunk of anything that you can touch fundamentally boils down to a collection of electric charges interacting with a heaving sea of zero-point scalar waves. Matter is not equivalent to energy, mass is energy that is constrained to a small space. More fundamentally, there is no mass, only charge. Charge is created by cosmic beings.

So you get more and more matter from innumerable active cosmic beings resulting in a growing universe; one that is extremely solid, a continuing chaos, unregulated, and accumulating without limits. It was set up once and then automatically, unregulated, keeps on growing. And when a cosmic being like you or me comes into this energy universe, he/she (spirits have no gender) is able to impose space and time on it, he/she can create space and time and he/she can place created energy in spaces and times which he/she creates. This is where you get spoon bending, spontaneous combustion, UFOs, ghosts, and paranormal phenomenon. Left is a farmer's field where a digital camera has caught some hydrogen standing waves left by a cosmic being (spirit). Spiritualists call this stuff ectoplasm.

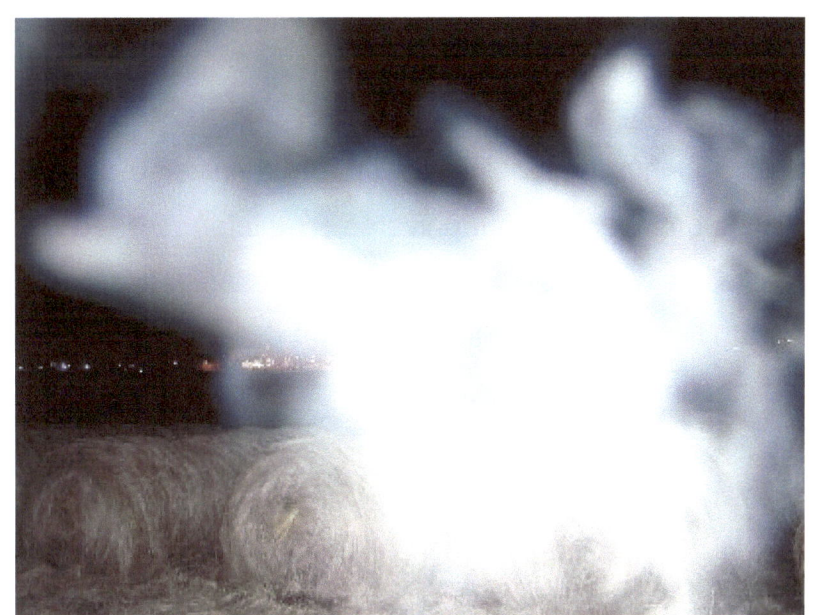

The accumulation of these space resonances produced by all the spirits in 500 trillion years or more and their conversion to heavy elements in stars has produced the Physical Universe as we know it today. Each galaxy is created by the beings that live in that sector. Energy that powers our sun and makes life possible on Earth comes from the beings in our vicinity. As we have seen, wave structures cannot escape the system they were created in unless they are converted into light (photons). Spirits are the source of cosmic mechanics. They produce matter, energy, space and time, but they are not composed of any of these. They are a nothingness, a zero. They are not composed of matter, energy, space or time so they have no size or form. They have no wavelength, mass or space, but they can create these. They are beingness without havingness, whereas we are beingness (humans) with havingness; i.e. bone, muscle, organs, skin, etc.

35) ZERO-POINT WAVES CARRY INFORMATION

There is more to the cosmos than vibrating strings and related quantum events. Life, mind and culture are part of the world's reality, and a genuine subject of Cosmic Mechanics should take them into account as well. Information is a real and effective feature of the Physical Universe. Although most of us think of information as data and what one knows, physicists are discovering that information extends far beyond the mind of an individual, far beyond the information stored in the orb, far beyond the information stored by all people. In fact, it is an inherent fact of nature. The great maverick physicist David Bohm called it "in-formation," meaning a message that actually forms the recipient. In-formation is not a human artifact. We don't create it by writing, speaking, calculating or emailing it. Ancient sages knew that in-formation is produced by the real world and is conveyed by the zero-point field in nature.

Our universe is not made up of vibrating strings, nor separate particles, but is instead constituted in the embrace of continuous fields and forces that carry information as well as energy.

This concept is important because the energy and information imbued "informed universe" is a meaningful universe. It helps us to understand ourselves and our universe. The informed universe is the most comprehensive concept of the world ever to come from science. It is a truly unified concept of life, mind and cosmos. When Dr. Milo Wolff discovered that there was no solid, discrete chunk of anything in the whole physical universe, he showed us the mechanism by which all parts of the Physical Universe communicate with all other parts.

This zero-point field informs all living things – the entire web of life. It also informs our consciousness. Intuitive people have always disagreed with scientific materialism which taught that we live in a world of inert, non-conscious matter moving randomly in passive space. Mainstream scientists defend the established theories of materialism. Maverick scientists at the cutting edge explore alternatives. For a while the alternative conceptions seem strange if not actually fantastic. Yet they are not the work of untrammeled imagination. They are based on rigorous reasoning and experiments. They are testable, capable of being confirmed or disproved by observation and experiment.

The Mechanics Of Paradigm Shifts

Investigating the anomalies that crop up in observation and experiment and inventing concepts that could account for them make up the nuts and bolts of fundamental research in science. If the anomalies persist despite the best efforts of mainstream scientists, and if a concept advanced by a maverick investigator gives a simpler and more logical explanation, a critical mass of scientists stop standing by the old paradigm. We have a paradigm shift. A new concept is then recognized as a valid scientific theory. Old concepts die hard, usually one funeral at a time. The young scientists, not resisting change for they have nothing invested, gladly embrace the new, more intuitive, theory. It may be some time before the "Big Bang" theory, the "man from mud" theory and the "particle theory of matter" are relegated to the scrap heap.

COSMIC MECHANICS

Scientific theories in a growing number of disciplines are turning more and more fabulous. We hear of dark matter, dark energy, multidimensional spaces in cosmology, with space resonances that are instantly connected throughout space-time by deeper levels of reality in quantum physics, with living matter that exhibits the coherence of quanta in biology, and in space- and time-independent transpersonal connections in consciousness research—to mention but a few of the new concepts in science.

The Mechanics Of Coherence

The principal kinds of anomalies that scientists confront today are anomalies of *coherence* and *correlation*. Coherence is a well known phenomenon in physics: in its ordinary form, it refers to light as being composed of light that has a constant difference in phase. Coherence means that phase relationships remain constant and processes and rhythms are harmonized. All parts of a system of such coherence are so correlated that what happens to one part also happens to all other parts.

Investigators in a growing number of scientific fields are encountering this surprising form of coherence, and the correlation that underlies it. These phenomena crop up in disciplines as diverse as quantum physics, cosmology, evolutionary biology, and consciousness research. They point toward a previously unknown form and level of unity in nature. The discovery of this unity is at the core of the next paradigm shift in science. It makes Cosmic Mechanics possible.

Scientific Remote Viewers are using this information in the zero-point field to solve crimes, find precious ore in the earth and discover the location of buried treasure. If you could think outside the box that your orb forces you to live in, you could have fabulous riches of knowledge, friends, family, finance, real estate and enlightenment. If you could know the winning Superlotto number the day before it is drawn, your financial worries would be over and your ethical, emotional, religious and spiritual dilemmas would be just starting. Here is an example of people who applied their skill and knowledge to generate fabulous riches.

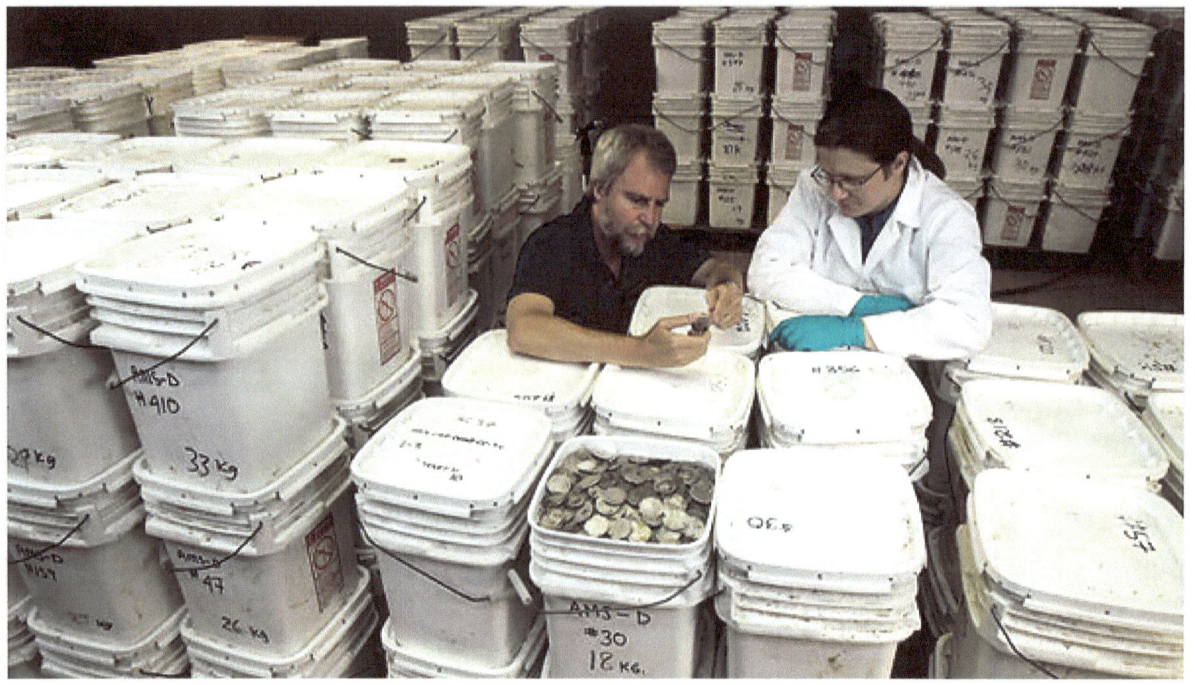

Odyssey Marine Exploration, via Associated Press

Greg Stemm, left, co-founder of Odyssey Marine Exploration, examining coins recovered from a shipwreck in the Atlantic Ocean.

MIAMI, May 18, 2007 — Explorers for a shipwreck exploration company based in Tampa said Friday that they had located a treasure estimated to be worth hundreds of millions of dollars in what may be the richest undersea treasure recovery to date.

Deep-ocean explorers for the company, Odyssey Marine Exploration, located more than 500,000 silver coins weighing more than 17 tons, along with hundreds of gold coins and other artifacts, in a Colonial-era shipwreck in an undisclosed location in the Atlantic Ocean, the company said in a statement.

The retail value of the silver coins ranges from a few hundred dollars to $4,000 each, with the gold coins having a higher value, the company said.

Citing security and legal concerns, Odyssey has not disclosed details about the discovery, including the origin of the coins and the identity or location of the site, dubbed Black Swan, but has said it is "beyond the territorial waters or legal jurisdiction of any country. The 6,000 silver coins that have so far been conserved are in "remarkable condition," Greg Stemm, the company's co-founder, said in the statement.

The find, which was announced on the same day that the publicly traded Odyssey held its annual stockholder meeting, came four years after the company found thousands of coins worth $75 million after excavating the Republic, a steamship lost in 1865 off Savannah, Ga. The company, which had reported losses for 2005 and 2006, saw its stock rise almost 81 percent to $8.32 by the time the market closed on Friday.

This year, Odyssey received permission from the Spanish government to resume a search that had been suspended on the wreck of the Sussex, a British warship that sank in the Mediterranean in 1694 with a cargo of coins that may be worth billions of dollars.

The company has not disclosed the methods or equipment it used in the Black Swan find. But it may very well include the aid of Remote Scientific Viewers. The largest documented previous find occurred in 1985, when the treasure hunter Mel Fisher found the Nuestra Señora de Atocha, a Spanish galleon that sank off the Florida Keys in 1622. The treasure included thousands of silver coins worth more than $400 million.

All this data is recorded in the zero-point field and you have the ability right now to access it. You could find untold billions in lost treasures, mother lodes, ancient artifacts and scientific secrets that will make you rich beyond measure. The only thing missing is your willingness to demonstrate that ability. All cosmic beings have the ability to know these things. But you won't demonstrate it because, if you did, it would mean that you would also know your own ancient crimes and hidden secrets. Everything would be revealed to you. You wouldn't be able to have that. Oh, no! Anything but that! That is too high a price to pay. It is much better that Greg Stemm of Odyssey Marine finds the treasure. Then you'll be able to sleep at night and have a life. You couldn't possibly face the past.

That's what Homo sapiens think. I hope my readers realize that anything can be forgiven. After all, the Physical Universe is only an illusion, and you cannot hurt a cosmic being. He's a nothingness! How can you hurt a nothingness? Only a being's own considerations could slow him down. So go ahead, confront what you've done and start a life! Get rid of your demon orb!

36) THE MECHANICS OF HYDROGEN

Hydrogen is the most common element in the Universe. It comprises about 99% of everything. This is because so many cosmic beings are producing it. It is a by-product of the activities of cosmic beings. Whenever beings experience stress, they produce ridges which manifest as hydrogen. Spiritualists call it ectoplasm and it is the causal agent in Human Spontaneous Combustion. It is formed by the union of an electron and a proton. It has no neutrons so it has very little mass. If you let some out of a tank in a laboratory it will rise up, flow out of the building and ascend above the earth's atmosphere to float in space. Hydrogen bonds so easily that it will stick to most anything and find its way back to a gravitational source such as a planet or sun.

The Mechanics Of The Orb

Spirits also create the little **21 gram ball of mass** that they take with them from lifetime to lifetime. It is composed primarily of hydrogen compounds and held morphogenically in place by zero-point fields. It is a highly sophisticated wireless quantum computer the likes of which mankind has yet to dream of. It contains painful recorded moments of excruciating loss and betrayal. It is the single source of all men's problems, stress, unhappiness, self doubt and his psycho-somatic ills. It is variously known as the reactive mind, unconscious mind or stimulus response mind. I'll show more pictures of this "orb" later. But this little ball does not contribute significantly to the mass of the universe. It does not burn or corrode or die. It was roughly 200 trillion years in the making.

The Mechanics Of Solar Energy

The mass of the universe is a by-product of the activities of cosmic beings. They produce ridges of hydrogen or ectoplasm which combines with other elements and falls into a star or sun. The resulting nuclear fusion and fission reactions convert it into heavier elements. Our scientists have determined that the chain reactions occur according to the following formula: $^1H + {}^1H \rightarrow {}^2H + e^+$ + neutrino. That means two protons (p^+) in two hydrogen atoms (^1H) react to form Deuterium (^2H = $1p^+$ & $1n$) plus a positron (e^+) and a neutrino. In the highly ionized stellar interior the positron will quickly "annihilate" with an electron ($e^+ + e^- \rightarrow$ 2 gamma-rays); the gamma-rays will be absorbed and re-emitted by the dense matter in the stellar interior, gradually diffusing outward and being "degraded" into photons of lower energy. When the gamma-ray energy reaches the photosphere (radiant surface of the sun) each gamma-ray will have been transformed into about 200,000 visible photons. The neutrino streams straight out of the Sun.

The next reaction is; $^2H + {}^1H \rightarrow {}^3He$ + gamma-ray, where the Deuterium (^2H) reacts with another hydrogen proton to form Tritium (^3He) or ($2p^+$ & $1n$) plus another gamma-ray. The first two reactions must happen twice to form two tritium (^3He) nuclei.

The third reaction is $^3He + {}^3He \rightarrow {}^4He + 2\ {}^1H$. Where the two ^3He nuclei react to form He ($2p^+$ & $2n$) giving two protons "change" plus a bit of added kinetic of the product nuclei. Note that the ^3He nuclei repel each other more strongly because they contain two positively charged protons. The initial reaction above can occur at temperatures as low as 1 million degrees Kelvin, but the last reaction can only occur at temperatures greater than about 10 million K. Further fusion reactions such as the carbon-nitrogen cycle, produce the heavier elements.

Supernovas collapse when their fusion-able fuel (hydrogen and helium) runs out. The expanding neutron core explodes with horrific force when it goes super critical. Super critical means that every neutron produced in the fissionable core produces more than one other neutron by

bombarding other heavy atoms. This produces a chain reaction which causes everything to blow apart with much more force than gravity can restrain. This process operates on the very same basic laws of nuclear physics used in nuclear power plants and atomic bombs! This is the basic source of energy and matter (condensed energy) in the Universe. Even Jupiter, (left) in our solar system, is so large that it is very hot and radiates heat into space due to the compression of its core by the megatons of its surface layers. Hydrogen is the primary component of Jupiter and the other gas giant planets. How do you suppose the hydrogen got there? Spirits have been creating it for trillions of years and it was accumulated by gravity into a huge gas ball.

Some people conceive this energy to be coming from God, but they don't know who God is. They don't even know themselves. They don't realize that they are a composite being, and they aren't their body. They can separate from their body and they routinely do this at death. Death is the transition to the universe of thought. Birth is the transition to the universe of matter, energy, space and time.

This postulated energy of the Sun was put here by cosmic beings. (Including us) This energy is eternally being created and will never get depleted, run down or be exhausted. This fact controverts the Law of Conservation of Energy. Where zero-point scalar waves are concerned, the great disciplinarian of the universes is the Minimum Amplitude Principle. This also holds true for electrostatic fields and magnets. The Sun and other stars are the conduit for all our energy. The energy of creation that we all produce goes through the Sun which makes photons that plant chlorophyll stores as sugars. We eat the fruit with the sugar and take into our bodies the photon's energy which we use for locomotion and body building. Even nuclear energy comes from a star because the uranium atoms used in nuclear energy were created in the fury of a super nova - a star exploding. This is very different from the energy that man uses to operate his body and machines. The fuel that man uses to operate his body and machines; food, petroleum, wood, coal and nuclear for instance, quickly gets depleted and must be replaced. Man is not being a source of energy. He is a spirit plus a body so he theoretically should be able to produce a never ending supply of energy. But he forgets who he is and pretends to be a body. To accomplish his purposes, man simply uses the domino-like cascading energy of the Physical Universe against itself.

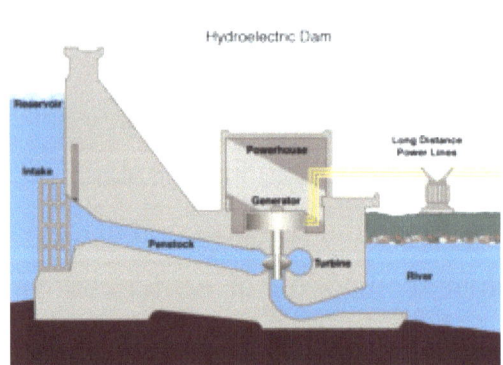

When man builds a hydro-electric dam which powers an electric drill a hundred miles away to make a hole in a piece of wood, he is just using the energy the beings have placed on Earth from the Sun. Our bodies are low temperature heat engines which run on light and chemicals. Man doesn't create his own energy; he just redirects the energy flows that are operating here on the surface of the Earth. For example, the Sun evaporates the water from the sea which condenses as rain over the mountains and flows into the dam, turning great turbines connected to electrical generators. Also, a great devastation on the Earth millions of years ago covered the rain forests with thousands of feet of debris. The heat and pressure working on the compounds in the rain forests, fish, dinosaurs and other animal's carcasses resulted in petroleum accumulations thousands of feet below the Earth's surface which we use for a source of fuel today. But the ultimate source of the physical energy we use today, of whatever variety, is the beings who set up the cosmic mechanics which powers this universe.

COSMIC MECHANICS

37) THE GHOST IN THE MACHINE

Now let's examine the cosmic beings that I said created our Universe. Are they God, the Creator? What is their potential? Do we have anything to fear from them? What evidence is there that they ever existed?

"This is a picture of my son, Philip, taken on Christmas night 2004. We were at my in-laws' house for dinner. After being up since the crack of dawn, Philip fell asleep in the middle of dinner. I took a picture and my sister-in-law also took one. She has not found her photo yet, but I know that she got the same bright light. There was nothing on that side of the room to create a flash. I do remember looking at my son thinking how beautiful and angelic he looked. I think it was an angel that we caught in this one." --Marla J. crazycatanddog@msn.com

What we know of God, we know very intimately, because that's us! We are the ghost in the machine. The life that beats in us and thinks in us and is in us has a direct connection to the infinite creative source. Do I believe in a supreme being? I certainly do, and there is a method of making more of them all the time!

God's first act of creation, expressed by the biblical statement, "Let there be Light," has validity in the material world. These statements make no sense to materialist scientists because they believe that light has to come from the sun or some other star. But now we see that God's light comes from beings existing in the zero-point field. We have pictures of beings creating this light.

So the cosmic beings who are responsible for the cosmic mechanics of this universe are us. We are the ones who long ago put the energy and time into this universe. We were the first ray gun and we are still putting out energy today, unknowingly.

We can put out waves to produce electrons that could burn infinitely more energetically than a laser beam. We are the ones who in present time are unknowingly creating the Physical Universe automatically by means of powerful postulates long since forgotten.

We are the source of the overwhelming power and force of this Universe. No one has ever allowed us to know how great we really are. We were only given inhibitions. We were told that we couldn't do things. "You can't fight City Hall!", "You can't go outside without your boots!", "You can't win'em all!", "Great expectations lead to great disappointments." and "You'll never amount to much." are some examples.

Creating quantum particles while dreaming (above). Notice that the 21 gram orb was pulled back into the man's head when the flash went off. The flash startled the man so he reactively pulled in the orb. The reactive mind, composed of space resonances, cannot be photographed when it is lodged in the cranium. Because a human body is host to a cosmic being and an orb, it is sometimes referred to derisively as a "container."

We are the Gods of our own universes. A universe is a whole system of created things. When you close your eyes and look with your mind's eye, you see a universe unfolding. This is your own universe, a completely different creation and in a different space than the Physical Universe. Although your own universe is in a different space than the Physical Universe, one exists right on top of the other. When you close your eyes and imagine yourself sitting in your car, touch the dashboard with your finger. Then open your eyes and notice that the point you touched in your mind is a point in the Physical Universe. No two objects can exist in the same space in the Physical Universe, but two universes can exist independently one right on top of the other.

The Physical Universe is common to all of us, but only you can put things into your own universe or take things out. The Law of Conservation of Energy does not apply to your own universe. You can create energy and objects without limit in your own universe. You can get excited about some project and get very creative, such as painting a masterpiece, building a spacecraft, designing a skyscraper, courting a spouse or directing the course of nations. Along with all this creation you produce space resonances which accumulate in the Physical Universe and, in large numbers, become subject to Newtonian Physics. In this universe there are no perpetual motion machines, it takes fuel to keep your car running, matter cannot be created nor destroyed, only converted according to Einstein's formula, $E=mc^2$. If we could see the Physical Universe as it really is, it would vanish, because it is only an illusion and when you see the way the magic was performed, it won't be real for you any longer. Others may still be spell-bound by the black enchantment, but you will be free. One of the lies used to keep the Physical Universe created is the propaganda that matter is composed of discrete particles. Getting billions of humans to agree with this illusion makes it <u>very</u> solid.

38) THE MECHANICS OF SPONTANEOUS HUMAN COMBUSTION

Spontaneous Human Combustion is the ability of the human body to blister or smoke or otherwise ignite in the absence of an external identifiable known source of ignition. In classic spontaneous human combustion the body burns itself more completely than can normally be achieved at a crematorium. The fires are internal in origin and result from the production of hydrogen gas by the cosmic being who is controlling the body. The hydrogen ignited by a cigarette, spark, or even body heat, burns at a very high temperature in atmospheric oxygen and consumes the body fats, tissues and bones completely. The being is usually under considerable stress when the combustion occurs. In other words, she/he's creating lots of gaseous hydrogen. If you've ever tested hydrogen in a high school chemistry lab, you know that it burns explosively at extremely high temperatures. Over the past 300 years, there have been more than 200 reports of persons burning to a crisp for no apparent reason.

In 1951 the Mary Reeser case captured the public interest in Spontaneous Human Combustion. Mrs. Reeser, 67, was found in her apartment on the morning of July 2, 1951, reduced to a pile of ashes, a skull, and a completely undamaged left foot. This event has become the foundation for many books on the subject of SHC since, the most notable being Michael Harrison's *Fire From Heaven*, printed in 1976. *Fire From Heaven* has become the standard reference work on Spontaneous Human Combustion.

In 1944 Peter Jones survived this experience and reported that there was no sensation of heat nor sighting of flames. He just saw smoke. He stated that he felt no pain.

Theories about Spontaneous Human Combustion
- Alcoholism - many Spontaneous Human Combustion victims have been alcoholics. (These were obviously beings under stress.) But experiments in the 19th century demonstrated that flesh impregnated with alcohol will not burn with the intense heat associated with Spontaneous Human Combustion.
- Deposits of flammable body fat - Many victims have been overweight - yet others have been skinny.
- Divine Intervention - Centuries ago people felt that the explosion was a sign from God of divine punishment.

COSMIC MECHANICS

- Build-up of static electricity - no known form of electrostatic discharge could cause a human to burst into flames.
- An explosive combination of chemicals can form in the digestive system - due to poor diet.
- Electrical fields that exist within the human body might be capable of 'short circuiting' somehow, that some sort of atomic chain reaction could generate tremendous internal heat.

No satisfactory explanation of Spontaneous Human Combustion has ever been given until now. The cosmic being's unknowing creation of highly volatile hydrogen gas is the cause of the body's cremation. Humans never realized that they had the ability to create matter and energy this way. They create Physical Universe matter and energy wherever they put their attention. But we have been confused by our university professors who teach scientific materialism. Hydrogen is estimated to make up more than 99% of all the atoms -- three quarters of the mass of the universe! This is because innumerable cosmic beings are producing it daily and the Physical Universe is accumulating it. This element is found in the stars, and plays an important part in powering the universe through both the proton-proton reaction and the carbon-nitrogen cycle. We are all helping to power this Physical Universe. We are all part of God, so let's give ourselves some credit! When you gaze in awe at the stars at night don't feel puny and insignificant like an ant! You are creating that Physical Universe and you are helping others to create it too. You can control the Physical Universe because you can start it, stop it or change it. You think that you cannot stop it because you've forgotten where you put your switch to turn it off. You think you cannot start it because you've forgotten the earlier universes that you have created. You think you cannot change it because you are no longer willing to use force and so you depend on the force of your body's engine or internal combustion engines to move things. You think the Atlantic Ocean is so big because it takes such a long time to swim across it. You measure everything compared to your puny body instead of compared to you, the cosmic being! You are huge without limits! You are infinitely larger than the whole Physical Universe! You have to start thinking big and thinking future! When you leave your head and move your viewpoint out a few hundred trillion light years you can look back to see the Physical Universe. My friends say it looks like a little brown oblong Sugar Daddy candy with points of light inside. You've forgotten how much you've contributed to the creation of our common universe. Over the eons you've created plenty of hydrogen. Stellar hydrogen fusion processes release massive amounts of energy by combining hydrogen atoms to form Helium. These are the same processes that are used to produce hydrogen bombs. There is no energy shortage on this planet.

The evidence that supports the theory that hydrogen is the cause of SHC is as follows:

- The body is normally more severely burned than one that has been caught in a normal fire.
- The burns are not distributed evenly over the body; the extremities are usually untouched by fire, whereas the torso usually suffers severe burning (around the solar plexus and heart chakras).
- In some cases the torso is completely destroyed, the bones being reduced completely to ash.
- Small portions of the body (an arm, a foot, maybe the head) remain unburned.
- Only objects immediately associated with the body have burned; the fire never spreads away from the body. SHC victims have burned up in bed without the sheets catching fire, clothing worn is often barely singed, and flammable materials only inches away remain untouched.
- A greasy soot deposit covers the ceiling and walls, usually stopping three to four feet above the floor.
- Objects above this three to four foot line show signs of heat damage (melted candles, cracked mirrors, etc.)
- Although temperatures of about 3,000 degrees Fahrenheit are normally required to char a body so thoroughly (crematoria, which usually operate in the neighborhood of 2,000 degrees, leave bone fragments which must be ground up by hand), frequently little or nothing around the victim is damaged, except perhaps the exact spot where the deceased ignited.
- In the witnessed combustions - people are actually seen by witnesses to explode into flame; most commonly. Here the witnesses agree that there was no possible source of ignition and/or that the flames were seen to erupt directly from the victim's skin. Unfortunately, most of the known cases of this type are poorly documented and basically unconfirmed. Sometimes there are no flames seen by the witness.

Non-fatal cases - Unfortunately, the victims of these events generally have no better idea of what happened to them than do the investigators; but the advantage to this grouping is that a survivor can confirm if an event had a simple explanation or not. Thus, there are far fewer cases of Spontaneous Human Combustion with survivors that can be explained away by skeptics without a second look.

Sometimes the victim will exhibit a mysterious smoke from the body. In these odd and rare occurrences smoke is seen to emanate from a person, with no associated fire or source of smoke other than the person's body.

39) ORBS; SMOKING GUN EVIDENCE OF IMMORTALITY

Why are we here on Earth? Do we just ride around the Sun for a few years and then its over? Its curtains? Where will we go after death? What will happen to us when we get there? Photographic pictures of paranormal entities can help us find answers to these perplexing questions. Seeing is the beginning of understanding. To complete the understanding process we must communicate with the entities, adjust our reality concerning them, and develop an affinity for them. Socrates, Aristotle, Bacon, Newton, Archimedes and Mendeleyev would have given anything to obtain this knowledge. This knowledge is priceless.

Dying makes all our earthly goals seem futile. Under hypnotic regression, people find their previous lives, but the healing process of finding their place in the spirit world is far more meaningful to people than finding their former lives on earth. Having a conscious knowledge of their soul life in the spirit world and a history of physical existences on other planets gives people a stronger sense of direction and energy for life. Discovering the significance of 21 gram orbs and their anatomy makes it possible for men to achieve total freedom. This book presents evidence that can give billions of people hope for survival on into infinity. As you will see, nearly all humans have orbs. It is the little collection of mental image pictures stored on ridges of space resonances that we make and take with us between bodies. It is our travel log of unconfrontable painful incidents. The space resonances of the orb are formed out of the quantum entanglement of waves produced by the human's universe and waves of the Physical Universe. They last forever just like

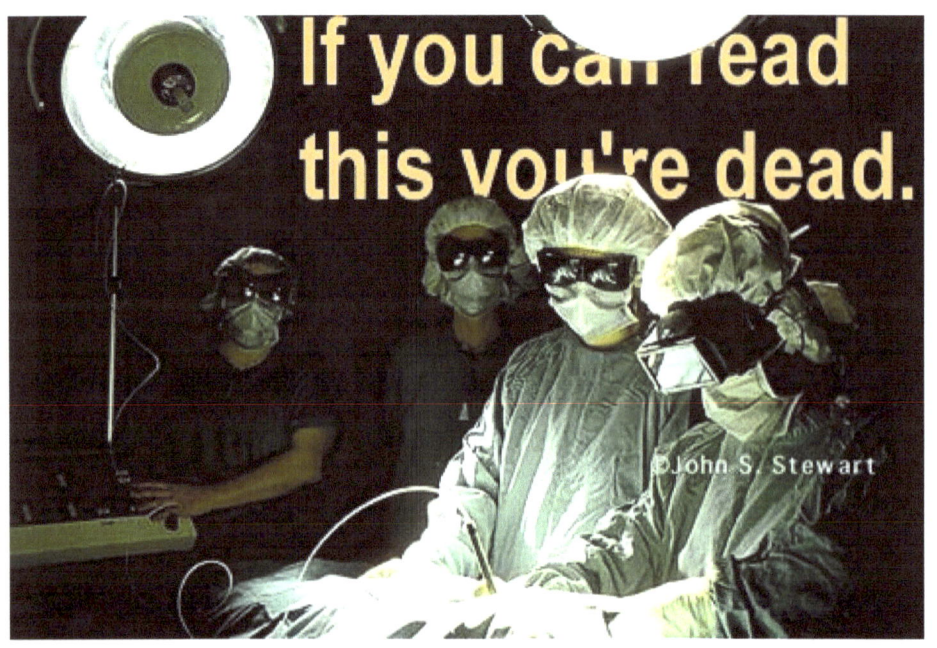

magnetism or the galaxy. They contain all our devastating pictures from our past lives. We only live once in a succession of bodies. The only questions left to determine our fate are: What will we do with our eternity? What state of mind will we create? Will we sabotage ourselves to an eternity of degradation, or will we be masters of all we survey?

We can decide what we are going to do this year. We could teach our dogs some new tricks, learn Fortran IV or get rid of our reactive minds. It depends on our value systems. It is our decision and our future depends on it. Will we continue to get weaker each lifetime until we have the ability of a roach? Or will we vanquish our orb this lifetime and take our place with the gods?

The goal of man is survival for the longest time and of the highest quality, but the goal of cosmic beings is to have the most interesting game possible, because their survival is assured. They are immortal. They are cosmic beings who take their creations of mental masses with them between bodies. If you knew that you were an immortal being, you would probably adopt new goals and purposes. This utterly heartless universe weaves a sticky web of forgetfulness and solidity. We are immersed in the mass of this universe and we've forgotten who we are. The little tract home in the suburbs with the white picket fence and 30 year mortgage may not seem as desirable when you find out who you really are. Your confidence in the American dream and the IRS method of financing it may be shaken. Your view of scientific materialism may take a beating when you see that man isn't made of the mud he thinks conceived him.

40) THE MECHANICS OF NEAR DEATH EXPERIENCES

When people have near death experiences and remember their past, they say things like, "I'm ejected out the top of my head. At the time of the accident I was pushed out. I'm a pinpoint of light, radiating. It's my energy. I look sort of transparent white. It's me, my soul, I am. I seem to grow a little as I move around. As I grow I look wispy...like a string...hanging. It's wonderful to feel so free, no pain, it's as if I'm suspended in air that isn't air. There are no limits. There is no gravity. I'm weightless. Nothing around me is a solid mass. There are no obstacles to bump into. I'm drifting."

So what is the essence, the élan vital that leaves a body at the time of death? It is the life-static, a nothingness, connected to a subtle 21 gram black mass that is usually transparent to human observers but which mysteriously shows up in digital photographs. For lack of a better term, people have been calling these things "orbs." The orb is the cause of all man's ills, wars, crimes, insanity and man's inhumanity to man. The orb is what keeps you small and powerless and the Physical Universe huge and powerful. It is as simple as that. If you could remember the power and glory you once possessed in the cosmos, would you settle for being human on this prison planet?

41) WHAT ARE 21 GRAM ORBS?

We have been building our orbs for roughly 200 trillion years. We had some excruciatingly painful experiences at first which overwhelmed us so we ridged up (flinched) and made a picture of them to evaluate later. That resulted in a little pin point of an orb. We had many more such horrible experiences which we recorded as space resonances in a little ball. The ball grew and grew until it now weighs around 21 ounces and fits nicely in the space around our bodies. During moments of injury it contracts inside our heads. Humans keep their orbs close to their bodies and take them along between lives. It is the reactive mind which is composed of unhandled space resonances. The resonances were produced by the interaction of violence in the Physical Universe and pain and unconsciousness in the cosmic being's universe. It took the shape of a ball because that is the shape of celestial bodies forming under the law of the Minimum Amplitude Principle (below).

TRAINING ROUTINES

In 1978 I was training to raise my confront of another person for two hours while sitting in a chair. After an hour and a half I got flu symptoms, fever, sweating and nausea. I fell off my chair. I thought a virus had infected me. When prodded to try it again the next night I got the same illness after an hour and forty-five minutes. I decided that I would tough it out and stay in my chair for five more minutes before falling on the floor. But to my surprise two minutes later all the flu symptoms vanished and I felt better than ever with a new powerful ability to confront people. My orb had made me sick! That's where my trouble was coming from. What a revelation! I would never have guessed that I had something in my mind that was sabotaging me. When a person starts recognizing that he has had a lot to do with creating his life, he starts to drop pieces of his orb. He achieves an ascendance over it. He is no longer oppressed by its awesome power.

COSMIC MECHANICS

Another incident of overcoming my orb was when I added an extra bath to my duplex in Los Angeles and did the plumbing myself. The tenants had to wait an hour for me to connect up the water main to the new bath. When I restored the water pressure by turning the water back on at the water meter I went under the house in the crawl space and found water spraying out from the union joint I had installed. This mess was very upsetting to me. I didn't want to disturb the tenants again so I grabbed a couple of large pipe wrenches and crawled in the mud under the gushing union joint to tighten it up. I tried to keep the water spray out of my eyes as I worked above my head with the wrenches. Soon the joint was tight and not leaking. I crawled out from under the house all muddy and felt 15 feet tall. I disagreed with the physical universe, made things go right, and exteriorized from under the house with more space. I had expanded my orb and overcome its constraint. When you can expand your orb and overcome its constraint so that it never bothers you again, you will have rehabilitated yourself as a cosmic being. You will be able to build your own universe.

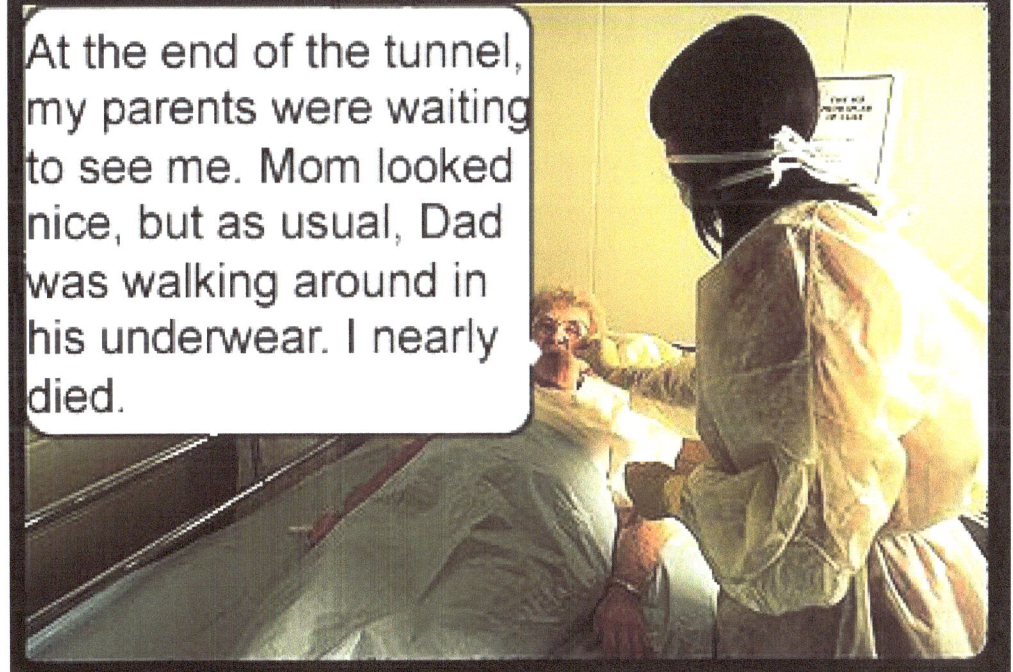

Although the orb masses are black, they appear white in the reflection of a photo flash. Our own full moon appears brilliant white in the clear night sky although it has the albedo (reflectivity) of bituminous coal. The new moon cannot be seen with the Sun behind it because it is as black as space. So, although it is a black mass, the Moon appears bright white in the glare of the Sun's energy.

Orbs most commonly appear in digital photos as a small ball of translucent gossamer material upon which can be seen ridges, spots and patterns. Some people, after some practice, can see the orbs with their naked eye in the photo flash of a camera. Rarely, orbs have been seen with normal vision to emerge from a person's head. This happened to my friend, Jacob, who was training to confront a lady for two hours near his home in Honolulu. After about an hour, sitting quietly, looking at the woman, her orb emerged from her head, wobbled around, and disappeared again back into her head.

The demon orbs vary widely in size. When not associated with a body, they are evidence of free spiritual beings that exist here unobserved along with rods and other beings of a very light density. It is important to stress here that these masses are not other-dimensional any more than humans come from another dimension. It is true that a human body is the interface between the world of thought and the world of matter. To that degree, orbs and humans are inter-dimensional, but the image obtained from a digital camera is totally a Physical Universe phenomenon. The shape, color, structure and energy of an orb are physical properties just as the shape, color, structure and energy of your favorite movie star are physical properties. They are just less dense and more refined microscopic wave structures. The ridges of the orb contain trillions of mental image pictures along with the decisions that are affecting the person today. If the person doesn't handle it, he/she will have a worse mental state next lifetime than he/she has today because the orb is growing and getting more powerful and destructive as the years and lifetimes go by.

I don't recommend hypnosis for therapy, but much has been learned about our essential nature from people in a deep state of hypnosis. One subject, as he recalled a death experience, exclaimed, "Oh, my god! I'm not really dead—am I? I mean, my body is dead—I can see it below me—but I'm floating … I can look down and see my body lying flat in the hospital bed. Everyone around me thinks I'm dead, but I'm not. I want to shout, hey, I'm not really dead! This is so incredible … the nurses are pulling a sheet over my head … but I'm still alive! It's strange, because my body is absolutely dead while I'm moving around it from above. I'm alive!"

42) THE MECHANICS OF HYPNOTIC INDUCTION

What happens if a hypnotist fails to bring a person out of a hypnotic trance? Whether or not the hypnotic induction is ended, the post-hypnotic suggestions will continue to be partially effective

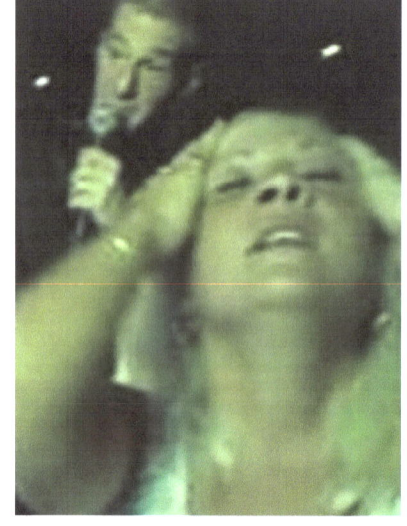

and make the subject more neurotic than he was before the induction. Anyone who volunteers for a stage hypnotist's show has got to be his own worst enemy. If he is brought out of the trance by the hypnotist and returned to normal wakefulness, what has he awakened to? He has awakened to the oppressive and debilitating hypnotic amnesia of his orb which leaves him practically unconscious. That is not a rational, awake state.

In 1993, Sharron Tabarn got up on stage to be hypnotized during a show. Towards the end of the show, she was told she'd receive 10,000 volts of electricity as she came out of hypnosis. Sharron suffered from a phobia of electricity, having received a major shock in childhood. Within a few hours of returning home she was found dead. Despite her mother's belief that the hypnotist was to blame, any links between hypnotism and her daughter's death were dismissed at the inquest.

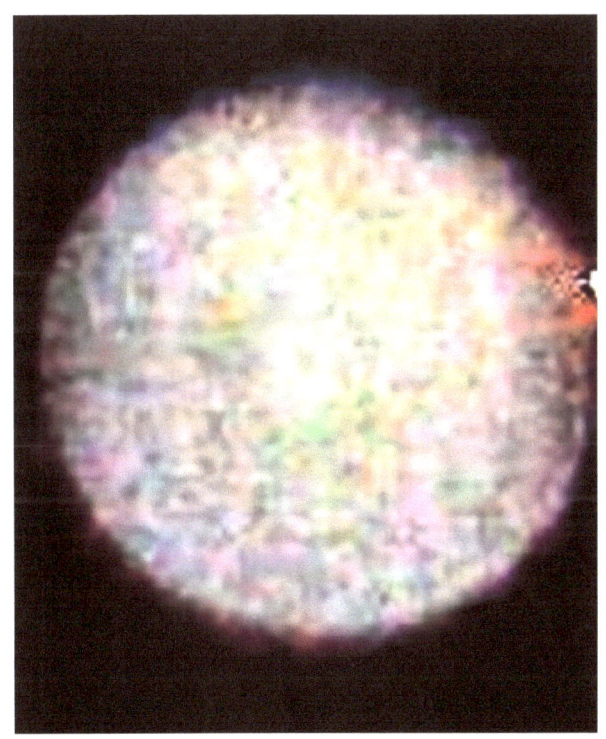

Popular writers such as Doctors Raymond Moody and Elisabeth Kubler-Ross have recorded the out-of-body near-death experiences of people severely injured in accidents. These people were considered clinically dead before doctors brought them back from the other side. Cosmic beings have the ability to leave and return to their bodies, particularly in life threatening situations when the body expires. These people tell of hovering over their bodies, especially in hospitals, watching doctors performing life-saving procedures on them. They report a euphoric sense of freedom and brightness around them. Some see brilliant whiteness totally surrounding them at the moment of death. While others observe the brightness in the distance while they are being pulled through a tunnel toward it. These memories fade with time after they resume life in their physical bodies.

Most of us have forgotten that we are cosmic beings capable of unlimited creative power and knowingness. We have each created a reactive mind (orb) to automate tasks and assist with our survival in the Physical Universe. But with the enslaving methods of oppressive beings, our orbs have been used against us. No one even suspects that our minds have been tampered with. Our brain is not our mind. Our reactive mind is the little 21 ounce ball of masses that shows up in our digital cameras, such as the one above. It has been enhanced to show the circuits and recordings that it contains. The mental image pictures and decisions are stored on the colored ridges. It is what we have taken with us for about two hundred trillion years, but it is not us. We are cosmic beings of unlimited creative power having no form, matter, energy, space or time. But we have the ability to produce these physical elements. We irresponsibly created our orbs and we've forgotten that we've done it, just like we've forgotten that we compulsively create the energy of this Universe and make the stars shine. But the orb is not us, just like our bodies and our cars are not us. They are our vehicles, our creations. The orb is the little ball of Physical Universe images and circuits that we have set up in the eons past to help us survive in this universe. But the device has gone awry and denies us our power. Men are all basically good, but that little ball of

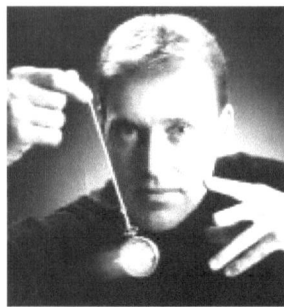

mental mass is the cause of man's inhumanity to man, insanity, criminality, illness and war. It is also the device used by oppressive beings and terrorists to control us and degrade us. It is the mechanism behind hypnotic control. In hypnotic rapport the hypnotist takes over the domination that the orb usually forces on the individual. It is the source of all the evil in the world. (I know that you'll think I'm over generalizing, but keep reading.)

Thus the hypnotist can distort your reality by making you sick, making your feet stick to the floor, making you speak Martian, making you think everybody in the room is naked, or by making you do the Michael Jackson Moon Walk when you've never done it before. Stage hypnotists do these tricks routinely. When the hypnotist's show is over, the subjects return to the distorted hypnotic reality forced off on them by their orbs. The orb exists both in your own universe and in the Physical Universe. It is the result of thwarted goals and the painful collision of your universe with the Physical Universe over time.

Professional counseling with live communication is required to eradicate this orb: to get rid of this insidious enemy and recover our infinite power and potential. By taking responsibility for your life and the creation of ridges in your orb, you can arrest its growth and power over you. If you are tough as nails you can overcome its influence and partially erase it if you conceive of yourself as a totally able being, unassailable in every direction and the master of your fate. (That is the ultimate truth! You are cause!) Then, whenever you have a doubt about your competence or get a feeling that you cannot do something, you'll recognize that that feeling is not you, but is coming from your orb. Accompanying this realization will be a reduction in the power of the orb to control you. You are, after all, a being of total cause, power, trust and faith who couldn't possibly be a victim because you are a nothing, a zero, in physical terms. You can only be controlled by the things you are attached to (like your car or your body your family or your orb) because it is impossible to hurt a nothingness and you'll never die. The important things in life aren't things; they are life-statics, nothings.

43) EARTHLIFE IS AN ELABORATELY CONTRIVED ILLUSION

Things aren't the way they seem for humans. It is like going to Universal Studios and taking the *Back to the Future* ride. In the middle of the show you can take your eyes off the screen while you are being hydraulically buffeted around in your "flying car" and you can see other people in their "cars" being buffeted around by the same computer program. It robs the experience of some of its frightening reality when you see that you are sitting on top of a set of hydraulic rams being jostled around while you watch lights and shadows dance on a screen in front of you. It is the same way in life. If you are human, you cannot be right. You are one of the "sheeple", a total effect. If you knew that the game of Earth Life was all a show and that you couldn't possibly get hurt, it would lose its excitement, its reality, its drama, its interest. If you had extended vision like the CCDs in digital cameras you couldn't be fooled. No, you'd be playing a bigger game.

You could be playing the warden of this planet busily hiding the smoking gun evidence of crashed disks, hypnotic implants, alien abductions, channeling, mind control, demonic possession, extra-terrestrials and artifacts observable on the moon and on other planets. You'd laugh at the antics of Homo sapiens "containers" riding Earth around the Sun year after year. You'd make sure to firmly establish a World Mental Health Organization and psychiatry which vehemently denies the existence of a world of spirits and you'd train your operatives in the methods of plausible denial to explain away evidence of the machinery of oppression. Oh, and by the way, you'd need a way to finance all this so you'd get the inmates to pay for it all by automatic withdrawals from their paychecks each month into a private corporation (IRS-Federal Reserve Bank) which prints money on paper and lends it to governments as if it were gold or real tangible property. But, more likely, you would be helping to salvage mankind from this awful tragedy of 21st Century human culture. It sure is easy to get paranoid in this line of investigation! You could get pseudo-paranoia; that's the fear that the IRS is out to get your money when in fact they are! Man's wars, his revolutions, his sufferings, all stem from his lack of data on the mind and spirit. With Psychiatric dominance of this field, we have had a century of untold suffering and violence. If you don't know how your mind works, you should find out fast, because oppressive beings that do know how it works are using it against you.

44) THE MECHANICS OF EVIL

All men are basically good, but they are ruled by an orb that turns them criminally insane. Between lives, the orb will make you feel like you are dying or worse. Imagine the plight of the psychiatric patient heavily overdosed with mind altering drugs. He has just had a fight with his family and killed three persons with his shotgun. He knows that there is only one way out; he puts the shotgun to his temple and pulls the trigger. Exterior, he looks at the mess he has created and thinks, "I just blew my brains out and I still want to die!" When he is reborn, he has forgotten about all this (thank God!) and he wants no one to remind him. He wants no hypnotic regression or mental therapy of any kind. Now he has a new game that he can be relatively successful at so he comes up to enthusiasm about his current condition and goes on to become President of Enron.

As people roller coaster through life their emotional tone rises and falls between Apathy and Enthusiasm. If you follow a spiritual path toward enlightenment you are going to lose a large portion of your orb and wind up exterior with full perception and in control of your life. At this point the domination of your orb will be negligible. Your goal is to become cause over matter, energy, space, time and life itself. You will fall short if you are irresponsible and allow anything; spirit, body, person or entity, to come between you and total cause.

Your life's spiritual goal consists of experiencing and stepping above whatever circumstances you find yourself in. Bad conditions don't just happen. The economic and cultural decay we are experiencing today isn't a random occurrence. It was caused. Unless we understand this we won't be able to defend ourselves against the unseen hand. Now that we have the ability to track the activities of orbs, we have another tool to secure our freedom. We have to use whatever resources we can muster to fight our way out of the trap of being a human body. It does no good to wish that our bodies were stronger or that our folks were richer. Our condition is what we have created. We made the choices that wound us up in this time and place. No one else is responsible for our condition. Once you totally accept that you, not fate, control your life, a portal opens silently within you, and without your realizing it at first, you begin a higher evolution. You take responsibility for your physical and mental creations and begin to recover your true identity and restore your capabilities of imagination, creativeness, thought, force and cause. These basics are nothing less than the foundation upon which rests all your hopes and dreams and any beauty you will ever see or feel.

I am cause. I cause things in the universe of thought between lives on Earth and I cause my own condition in mortality. You were the one who chose your family and your social conditions. On Earth it takes longer to see the effects of your own cause because objects are so heavy and dense here. Things move very slowly. For example, when you poison your body with cigarettes, alcohol and drugs, you don't get really sick for a couple of years. It takes a while for the effect to manifest. Similarly, when you save money and work to create your dreams, it takes a while before your dreams manifest. But you always get what you put your attention on. So be smart about what you put your attention on. When we cross over to the other side as a cosmic being, all we will be able to take with us will be our travel log pictures and our state of mind. Put yours in excellent shape. **There is no therapy on the other side to rid you of your orb.** You need a body and the technology and someone to help you vanquish your orb now, on this side!

INHUMAN BEINGS - The Chemicalized Personality

A mother murders her five children. High school students massacre their classmates. An Iraq vet stabs his wife 71 times. How can this happen?

A common thread in these occurrences is the fact that the killers have been taking psychiatric medications.

But that is too simple. So we hear about "post-partum depression" and "combat stress." In the case of the teens, it's "the breakdown of the family" or it's the music, the movies, the video games.

The real answer is the dehumanizing effect of drugs and the beings who distribute them.

A human being has more than one aspect. There is a definite electro-chemical component. The body physically functions via electro-chemical processes and biophotonic communication.

Then there is that aspect which perceives and reasons and creates, the cosmic being. This is not electro-chemical. When people communicate with each other, it is not chemical molecules that are exchanging ideas. This is the spiritual aspect; the conscious, aware individual.

There is also a mental component—a mind—which is an interactive link between the reasoning factor and the physical.

A healthy mind (motivated by the spirit) is analytical.

A less healthy mind is less analytical and more and more reactive. It operates on a stimulus/response basis, motivated by random factors. A troubled, unhealthy mind doesn't reason. It doesn't perceive well. It reacts to stimuli. This is the 21 gram orb.

For a long time now, the mental health establishment has been telling us that we are chemical in nature. They would have us believe that they can solve our problems with mood-altering drugs—a little dash of this and a pinch of that.

That approach may work at the purely physical level, as in taking antibiotics to handle infection, but it is not the physical component that gives us our rationality, our humanity. It is not the molecules in the brain that are thinking and perceiving, loving and caring, creating great music and poetry.

No, the physical component is comprised of cells and electrical impulses, which are as reasoning and creative as an avocado or the electric current that powers your toaster.

When a person is troubled, he is already sliding in the direction of the reactive, unthinking, physical impulses of his/her orb. To then give him/her chemical, mood-altering drugs, pushes him further in that direction. While the sedative effect may appear to calm him down, he is becoming, more and more chemicalized.

So is it any wonder that these killers seem less than human? They ARE less than human. Though they can appear bright and calculating at times, real judgment is gone. They are completely reactive; uncaring, alienated.

Their minds bubble and boil like the mass of chemicals they have become. The analytical capacity is gone. The spirit is gone. Their humanity is gone. They respond randomly and literally to stimuli (such as music, movies and video games). This is the condition the psychiatrist has created to install the hypnotic command to murder their classmates and themselves. Evil psychiatrists don't want their victims to know who they are and how their orb works.

Then their victims lash out with violence at the imagined demons and enemies in their own unreal world. They have been mentally short-circuited by the drugs that are supposed to be helping them and hypnotized with delusions. It is the ultimate betrayal.

And when their bizarre, chemically induced, nightmare world collides with the world of OUR reality—which consists of living people, loving families, children, teachers, learning, accomplishment—a slaughter ensues and we are left to wonder "WHY?" "WHAT HAPPENED?"

The Answer: psychiatry happened. And why would anyone perpetrate such a crime as to drug children and adults, driving them insane, all in the name of help? It's too horribly simple. It's a multi-billion dollar business.

The good news is that when society wakes up to these facts, we will cease to allow these evils to occur. It's time to wake up.

The Virginia Tech massacre was not "caused" by a drug.

Drugs don't cause things; they just beef up people's orbs, lower their awareness, and confuse them. The Virginia Tech massacre was caused by one or more psychiatrists who implanted the shooter with hatred and a command to kill. We could probably find out who did it by locating the most "prominent" psychs in the area and see which one went on "vacation" on the day of the shooting or just before. Look there for the "being," not a drug. Beings manufacture "Manchurian Candidates," the drugs don't. The drugs just make people sick and insane. The U.S. Congress has yet to investigate the role of psychiatric drugs relating to school shootings despite international drug regulators warning these drugs can cause mania, psychosis, hallucinations, suicide and homicide.

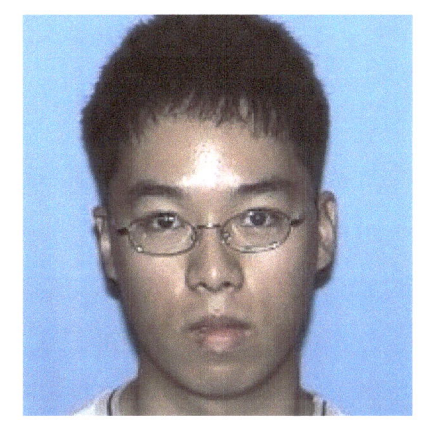

Cho Seung-Hui's murderous rampage – during which he killed 32 students and faculty members at Virginia Tech – is prompting research into gun laws, resident aliens and graphically violent writings. Investigators also may want to check his medicine cabinet, because psychiatric drugs have been linked to hundreds of violent episodes, including most of the school shootings in the last two decades.

The New York Times has reported the killer was on a prescription medication, and authorities have said he was confined briefly several years ago for a mental episode. They also have confirmed that the "prescription drugs" found among his effects related to the treatment of psychological problems.

Reviewing most of the mass murders from the late 1800's through the mid-1990s, the evidence is overwhelming that such events don't happen on their own. There is a psychiatrist with his shock machine, pharmacy, ice-pick, straps, handcuffs, usually an assistant and an overt or covert but always complex and continuous determination to harm or destroy. With blood on his hands he proceeded to torture and drug and hypnotize the shooter into an utter state of confusion and psychosis.

You can count on that, just as surely as his "patient" will show up on command with ammo, weapons, bombs or other violent means of destruction to take the lives of as many children or others as he or she can. Sometimes it's a mother who drowns her five children or throws three of them from a bridge and jumps in behind them. It is always violent.

The Mechanics Of Crimes Committed Under Hypnosis

Persons under hypnosis commit heinous crimes. In March 1951, in Copenhagen, Denmark, a 30-year-old man by the name of **Palle Hardrup** walked into a bank with a gun and demanded money. He killed a guard in the process, and later claimed that he had been instructed to do so by the hypnotist, Bjorn Nielson. Nielson eventually confessed to having engineered the crime as a test of his hypnotic abilities. One of the first tasks was arranging for Hardrup's girlfriend to have sex with the hypnotist. The other tasks were robbery and murder. Nielson convinced his victim that the robbery funds were to be used for worthwhile political purposes. The end, Hardrup was told, justified the means.

COSMIC MECHANICS

The Charles Whitman mass murder in 1966 at the University of Texas in Austin is similar to the Virginia Tech massacre. You may have heard that the Virginia Tech massacre surpassed the previous "record" set and held by Charles Whitman since 1966. Whitman killed 16 outright and a seventeenth victim died later of shooting related injuries. Whitman also wounded 31 people. The Charles Whitman "Texas Tower" shootings, sometimes called the "first" school mass murder of the modern era, gives us clues as to what is at stake today: Not only are these incidents psychiatrist induced but indeed, are likely fallout of the ubiquitous psychiatric experimentation done on school campuses.

As you may know, the files of the Grand Jury that looked into the University of Texas, "Texas Tower" massacre by Charles Whitman back in 1966 were sealed at the request of the FBI and were never to be revealed.

However, the local Austin newspaper used the Freedom of Information Act over twenty years later to crack the sealed documents for an anniversary story on that mass murder. It turns out that the FBI had testified privately before the Grand Jury, after having publicly stating two days after the incident that, "there were no drugs involved."

However, as reported by the Austin American Statesman, behind closed doors the FBI revealed that, "Charles Whitman had been "eating amphetamines like popcorn." The data I have is that Charles Whitman was part of a secret, Navy funded study, being done on the UTA campus by the campus psychiatrist at the time, Maurice Dean Heatly who prescribed Dexedrine (amphetamine) on top of a Valium prescription by the school medical doctor, Jan Cochran.

Most people have also forgotten that the night before he went up into the Texas Tower on the campus, he had brutally stabbed, slashed and shot his pregnant mother and his wife to death.

So, Charles Whitman's and Maurice Dean Heatly's now infamous record for mass murder was just broken by this new assassin at Virginia Tech. The first to know that Whitman was exhibiting a classic, literally text book, psychiatrist induced amphetamine psychosis, was Heatly, the psychiatrist who treated him and wrote the prescription. Heatly was an accomplice, before, during and after the fact.

My opinion is that these school and other mass murders of the modern era are at least 98% provable to be strictly a phenomena of psychiatric induced mass murder. These individuals don't just go mad, not to this extent, to kill others and themselves. They are driven this mad by the deliberate if not negligent and irresponsible use of prescription psychiatric drugs, their "counseling" and electric shocks.

The mass murders we are seeing in America, the multiple homicides and other murders and suicides, very much parallel the homicide bombers and other terrorists that have appeared in recent years in the Middle East, recruited, "counseled" and sent with bodies full of the drug Ecstasy (Methylenedioxymethamphetamine, better known as MDMA) or some other form of "speed" to carry out the directions of their cell leaders to kill as many people as possible, without hint or warning.

Of special note is the fact that there are presently over 3,000 amphetamine-type drugs, all fully developed for the psychiatric industry to use and logged with the FDA, all awaiting approval for "human trials," most of which trials will be conducted on our school campuses, using our children as the human subjects of those experiments.

Mass murderers are manufactured by people. Those people then sent them out among us to do their work. The drugs, like ammo or a bomb vest are just part of the package, like the clothes they wear.

Further, recent laws, under the guise of "patient confidentiality," were created by the psychiatric industry to protect the psychiatrists who send terrorists, against exposure. It seems it is illegal today to expose the acts of the psychiatrist. He's protected, not his victims.

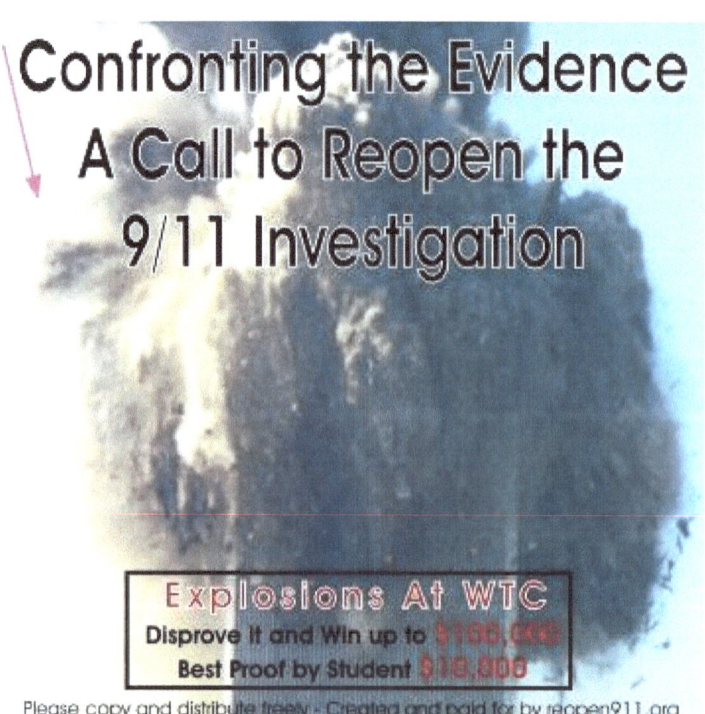

Our planet is in a world of hurt. Multi-national corporations are raping the earth, defrauding their investors and cheating their employees out of their retirement benefits. Jeffrey K. Skilling was sentenced to 24 years and 4 months in prison for his role in the pervasive fraud and conspiracy that led to the bankruptcy of Enron. The morals and ethics of our presidents and congressmen are deplorable. After 9/11, President George W.

COSMIC MECHANICS

Bush took us to war in Iraq based on the false assertion that Sadam Hussein was building weapons of mass destruction. But the real destroyers of the World Trade Towers have been allowed to escape judgment for their crimes. When we measure the possibilities created by 9/11 against what we have actually accomplished, it is clear that we have found one way after another to compound the tragedy. Homeland security is half-finished, the development at ground zero barely begun. The war against terror we meant to fight in Afghanistan is at best stuck in neutral, with the Taliban resurgent and the best economic news involving a bumper crop of opium. Iraq, which had nothing to do with 9/11 when it was invaded, is now a breeding ground for a new generation of terrorists. There is overwhelming evidence that the 9/11 Commission white-washed the actual evidence of the destruction of the World Trade Towers because it was an inside job. Aircraft striking the towers did not bring them down, demolition charges did. History repeats itself: the sinking of the cruise liner Lusitania by a German U-boat that was given false information that it carried war munitions brought America into WWII in Europe. Getting out of the Japanese way allowed them to bomb Pearl Harbor and bring Americans into the war in the Pacific. In a purely artificial chess game Roosevelt sacrificed over 2400 American Seamen's lives, thanks to his power as Commander in Chief of the Armed Forces. By over-looking the obvious facts of an attack by Japan on Pearl Harbor, Roosevelt was able to control both the political and economic systems of the United States. Most of American society before the Pearl Harbor bombing believed in the idea of isolationism.

Franklin D. Roosevelt knew this, and knew the only way in which United States countrymen would take arms and fight in Europe's War was to be an overt action against the United States by a member of the Axis Power. There are numerous accounts of actions by Roosevelt and his top armed forces advisors, which reveal they were not only aware of an attack by Japan, but also they were planning on it, and instigating that attack. Roosevelt gave Japan a sign of United States military preparedness and threat of attack. He also put strong controls on Japans trade and economy. He insisted that the Dutch refuse to grant Japanese demands for oil and a

complete embargo of all trade with Japan by the United States. The Chinese government of Chiang Kai-shek was completely against Japan, and with economic support from the United States, they were able to deny certain possessions from Japan. By provoking the Japanese and the foreknowledge of an attack on Pearl Harbor, Roosevelt along with his top advisors and the Federal Government are truly to blame for the loss of American life and property.

Public opinion is manipulated this way by cosmic beings overwhelmed by their orbs. Their method is to create a problem, have the public react to it and then offer the subversive solution. This is how the reptilian "custodians" have pushed through most all of our oppressive laws. They have a very low opinion of humans. They justify their crimes by thinking that humans with bad instincts are more in number than the good, and therefore the best results in governing them are attained by violence and terrorism, and not by academic reasoning. Reptiles transfer their values to humans, thinking that every human aims at power, every human would like to become a dictator if only he could, and rare indeed are the humans who would not be willing to sacrifice the welfare of all for the sake of securing their own welfare. This is the reptilian thinking. Thus the most insidious of all forms of torture is used, mental rape. It is perfect because you never know what hit you. You have no target to strike back against. You feel like a bull trying to attack a pesky matador, only to get fooled by the red cape. The bull never gets the satisfaction of really goring something solid with its horns so he's dragged out of the arena unhurt but totally demoralized.

To give you an idea of how debased some orb controlled minds can get, here is a translation of a speech by Professor Beria In 1947 at the Lenin University auditorium in Leningrad, Russia. Professor Beria addressed a small group of American Communist students who had come to study Psycho-politics.

* * * * * * * *

(The translation of the lecture from the Russian viewpoint is given as follows): "Psycho-politics is the art and science of asserting and maintaining dominion over the thoughts and loyalties of individuals, officers, bureaus, and masses, and the effecting of the conquest of enemy nations through "mental healing." It is an important but less well known division of Geo-politics. It is more esoteric because only the elite, highly educated strata of "mental healing" experts deal with it. Psycho-politics is what we use to effectively reach our goals. The most important first step in our program is to produce the maximum chaos possible in the culture of the enemy. We utilize chaos, distrust, economic depression and scientific turmoil. Then a weary capitalist populace will seek refuge in our Communist State because only Communism can resolve the problems of the masses.

COSMIC MECHANICS

A psycho-politician must diligently produce the maximum chaos in the fields of "mental healing." He must recruit, promote, and use all the agencies, personnel and facilities of "mental healing" until the entire field of mental science is dominated by Communist ideals. The promise of unlimited sexual opportunities, the promise of complete dominion over the bodies and minds of helpless patients, the promise of complete lawlessness without detection, can thus attract to "mental healing," many desirable recruits who will willingly fall in line with psycho-political activities. To achieve these goals the psycho-politician must crush every "home grown" variety of mental healing in America.

The actual teachings of Freud, James, Eddy and others amongst your misguided people must be discredited, defamed, arrested, and stamped upon by the American government until there is no credit in them and only Communist-oriented "healing" remains. The greater the number of insane persons in a country, the greater is our control and facilities. As the insanity increases, an atmosphere of emergency is created which justifies our use of violent treatments such as electro-shock, prefrontal lobotomy, trans-orbital leucotomy, and other operations long-since practiced in Russia on political prisoners.

You must dominate the minds and bodies of every important person in your nation. You must achieve such authority over the pronouncement of insanity that not one statesman so labeled could again be given credence by his people. You must work until suicide arising from mental imbalance is common and elicits no general investigation or remark. You must ensure that every teacher of psychology knowingly or unknowingly teaches only Communist doctrine under the guise of "psychology."

The institutions for the insane in your country are prisons which can hold a million persons without civil rights or any hope of freedom. You can practice shock and surgery on these people so that they will never again draw a sane breath. These treatments must be made common and accepted, and you must sweep aside any group seeking to treat by effective means, as all our actions and researches could be undone.

Dr. F. Lee Aeilts

You must dominate the fields of psychiatry and psychology as respected men in hospitals and universities. Society should expect every recalcitrant young man to be brought into court and be assigned to a psycho-political operative, be given electric shocks, and be reduced into unimaginative docility for the remainder of his days. Movements to improve youth should be invaded and corrupted, as this might interrupt campaigns to produce in youth delinquency, addiction, drunkenness, and sexual promiscuity.

With psycho-politics you can crush your enemies as if they were insects. You can step on them and put them out of their misery. You can cripple the efficiency of leaders by striking insanity into their families through the use of drugs. You can defeat them by testifying to their insanity. You can actually turn them into blathering idiots with our electroshock and neurosurgery.

The actual simplicity of the subject of pain-drug hypnosis, the use of electric shock, drugs and insanity-producing injections should be masked entirely by technical nomenclature, by the protest of benefit to the patient, by an authoritarian pose and position, and by carefully cultivating governmental positions in the country to be conquered.

You can change forever the loyalty of a soldier or a statesman or an opinion leader or destroy their minds but there are certain dangers. It may happen that remedies for our "treatments" may be discovered or a public hue and cry may arise against "mental healing." As a result, all mental healing might be taken out of our hands and be given to ministers. The Capitalist thirst for control, Capitalistic inhumanity, and a general public terror of insanity will guard against this. But should independent researchers actually discover means to undo psycho-political procedures, you must not rest or sleep or eat or withhold the slightest bit of available funds to campaign against it, discredit it, strike it down and render it void. These must be defamed and excluded as "untrained," "unskillful," "quacks," "perpetrators of hoaxes" or dangerous. The presence of an effective mental health technology could undo all our actions and researches. It must continually be stressed to mental health groups that the entire subject of mental health is so complex that none of them, certainly, could understand any part of it.

Use the courts, use the judges, use the Constitution, and use its medical societies and its laws to further our ends. Diligently work to dominate the field of "mental healing" so that you can pass your own legislation at will, carefully organize all healing societies, constantly campaign about the terrors of society and by pretense of effectiveness, make the capitalist, himself, by his own appropriations, finance a large part of the quiet Communist conquest of the nations.

No Russian agent could be even remotely effective without a thorough grounding in Psycho-politics. You carry forward with you a Russian trust to use well what you are learning here. By psycho-politics create chaos. Leave a nation leaderless. Kill our enemies and bring to Earth, through Communism, the greatest peace man has ever known. Thank you."

Some cosmic beings have orbs with limitless ambitions, burning greediness, merciless vengeance, hatreds and malice. It is from them that the all-engulfing terror proceeds. What human will ever suspect that they were stage-managed by such beings according to a political agenda which no one has so much as guessed at in the course of many centuries?

45) WHY CAN'T WE SEE ORBS WITH OUR EYES?

Some humans have trained their eyes to see orbs briefly, but most cannot. Most don't even believe that orbs exist. Humans have a very heavy density or mass. We over-value the physical and under-value the spiritual and mental side of life. We seem to be composed of just a little bit of spirit and a lot of mass in our animal bodies. We are extremely heavy and slow compared to other types of life forms. In the Physical Universe things move agonizingly slow compared to the speed of things in the thought universe. Our eyes do not perceive the long wavelengths of the infrared range the orbs reflect so they go unnoticed as disembodied beings watch and move around us. There is a whole universe of unseen beings around us watching and doing what they can to further their agenda and influence the course of human evolution. They alter our minds (orbs) between lifetimes.

Humans take their orbs with them everywhere they go. Orbs are clear so that we can see right through them. They are made up of standing waves that reflect the light from a flash camera and show up well at night when the flash goes off. (For clarification, review the section on wave reflection when the density of the wave medium varies.)

Above, is a picture produced from a pinhole camera in 1896. The photographic emulsion was sensitive to the orb between the farmer's boots. Just as the Physical Universe, from one side to the other, is a ball of twinkling lights, so also is an orb a miniature universe of the being who created it. The difference is in the size and power. The physical universe is huge and powerful compared to the orb universe.

46) THE MECHANICS OF OUR PRISON PLANET

On this prison planet none of us escapes the reach of the tax man. We are each conned into believing we work for ourselves and that the taxes we pay go into administrating our government, building the infrastructure and defending our nation. But I think that a portion of the taxes we pay go to the world bank and are used for some nefarious psychiatric or extraterrestrial purpose contrary to our best interests. Note that the orbs works contrary to our best interests and so does the Physical Universe. Mankind is unaware of the existence and the oppressive influence of disembodied beings with orbs. Several internet groups exchange pictures of the orbs that appear in their photos but they have no idea of their significance. Their covert influence and insidious domination makes them the ideal weapon of a degraded being. It keeps us pinned down, unconscious and out of the game.

Here are five frames from a night vision camcorder that caught an orb in motion. The first frame shows the trail moving left to upper right and then it flies down in the second frame. The third frame shows it moving away from the camera and turning around a column before exiting to the right. Evidently the orb flew through the column because the last frame shows the remnant of light on the near side of the column.

COSMIC MECHANICS

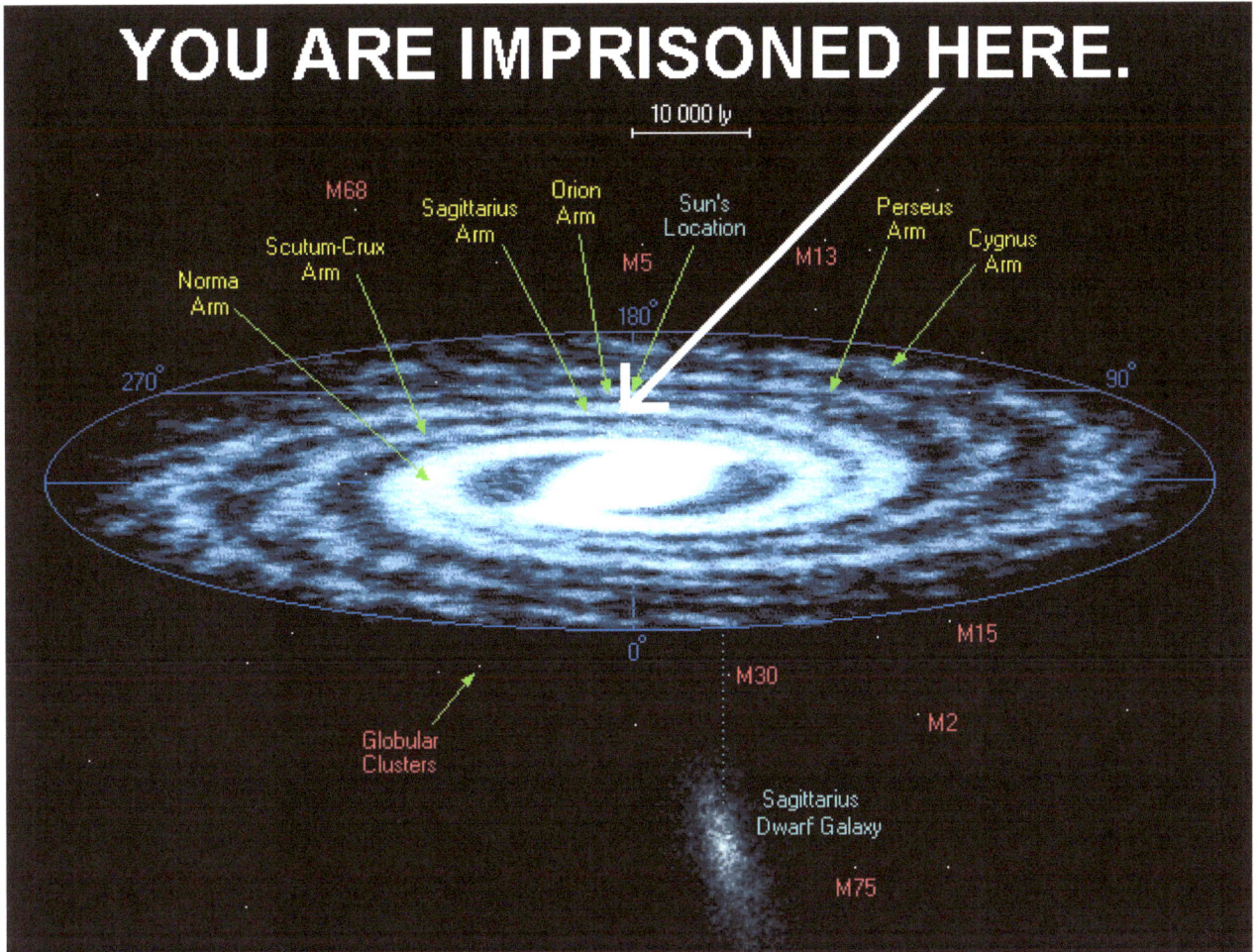

How can this little orb wreak such havoc? Note that the Federal Reserve Bank has never been audited and the public will never find out who owns it. The pharmaceutical industry is dumbing our kids and poisoning our people. Psychiatry is electro-shocking people, drugging them, cutting out their brains and ruining their lives. The US government is creating war in Iraq, not stopping it. What does this manifest evil have to do with the 21 gram orb? It is the single source of all the problems, unhappiness, stress and psychosomatic illness that we have in the world. Hollywood teaches us that there are good men and bad men in this world, but this is not true. All men are basically good, but they are dominated by their orbs which cause them to act like insane criminals. Even the reptilian "custodians" have orbs.

Everything that survives in this universe has to pay a price. In the wild it's the survival of the fittest. The rest are eaten to support the fittest. Trees offer shade or fruit. Horses offer labor, pigs offer the meat of their bodies and the people of planet Earth are paying their way or they wouldn't be allowed to survive. Each person on the planet contributes to Earth's gross global product and a small part of their taxes go to the World Bank for distribution to the reptilian "custodians." I must admit that this system is vastly superior to the slavery of ancient Egypt where we were each forced to work under threats of excruciating pain and death. But in those days we knew who the oppressor was. When 20 slaves couldn't pull a monolith up a slope, two were killed by the task master, and amazingly, the remaining 18 managed to move the stone. It was all necessity level.

In the last 10,000 years the minds of men have been "fixed" and the culture has been set up to secretly steal the fruits of the labors of man. The parasites are invisible. If we didn't have materialistic psychiatrists who drug us, scientific materialism, and all types of false data abounding on the Earth, the reptilian "custodians" couldn't keep us working for their interests. If we could see the spiritual beings surrounding us we wouldn't believe the university professors and the psychiatrists for we would know that we are spiritual beings inhabiting physical bodies. We would know that we are being conned. We would know that we are powerful beings who could create a much better world of our own and not have to be an unknowing slave to anyone. This universe is like a spider web. The more you struggle in it the tighter the grip.

Nearly every man, woman and child on this planet is in a hypnotic trance insanely and slavishly producing products for their captors, their hypnotic operators. And they don't suspect a thing! Tell any one of them that they have an insidious, self-defeating orb that is doing them in and they will vehemently deny it and tell you that they feel just fine and..."don't need any help, thank you." Their minds have been imprinted with other people's experiences and contra-survival urges and they can't tell which of their memories are their own and which are someone else's. All hostile, angry behavior is a product of this imprinted mind which is dedicated to protecting the individuated self against others who are viewed as outsiders.

I know this will seem incredible to you but an antagonistic man is forced by his own orb to reflect back to you, antagonistically, all communication you send to him. He cannot do anything else. His orb has powerful circuits which overwhelm him. I mean the man is powerless to do anything but watch his own mental circuits bounce back everything said or done to him. He does not have the ability to give a fresh analysis to each statement given to him. No, he just reacts out of control and then justifies his senseless acts, not realizing that he is not the one making the decisions. If you tell him that he is criminally vandalizing cars by throwing stones at them he'll call you a criminal. His orb, which has been tampered with, does the dramatizations automatically out of the quantum mechanical interference patterns programmed into it   He can't tell the difference between his thoughts and actions and his orbs commands. His awareness is extremely low. He is not aware of his dependence on others, on groups, on his country, on mankind, on the physical universe, on spirits, and on God. He thinks he is one man against the universe when he should be on a team with others creating an ideal universe with the help of groups, countries, spirits and God. There are no outsiders, which every being could understand if they could only see themselves from an exterior view. We are all part of the brotherhood of man, all spiritual beings of unlimited potential.

Humans are a part of the whole that we call the Physical Universe, a part limited in time and space. They experience themselves, their thoughts and feelings, as something separated from the rest--a kind of optical illusion of their consciousness. Their minds create this illusion, and we know it has to be an illusion because our scientists have shown us that there is nothing solid in this universe. They have not found a real, solid particle. Even the atoms are found to only be composed of agitations within agitations. No solid atomic particle has ever been found by our physicists. Objects in our environment are only solid because we say they are. Our belief system makes them solid. You have to understand that the images that enter our eyes are upside down on our retina. So our minds interpret them by flipping them over, right side up. The images are also only two dimensional, so our minds create a three dimensional construct to interpret them. Then our belief system and orbs modify our sensory inputs to create the illusion we experience as reality. **We create reality with zero-point scalar waves.**

Just visit a stage hypnotist show sometime to see the extent that our belief systems distort our reality. The hypnotist will give a participant a post-hypnotic suggestion that he cannot see his daughter. Then the hypnotist will wake up the man and his daughter will come up on stage from the audience. She'll stand directly in front of her hypnotized father so that he is looking directly at her stomach. The hypnotist will ask the father if he sees his daughter and the response will be, "No." Then the hypnotist will hold his watch up against the girl's spine and ask the man what he is holding. The man will say, "A watch." The hypnotist will ask the man if he can read the serial number of the watch and he'll say, "Yes, it is 45836459." The reality of the hypnotized man was distorted so much that he couldn't see his daughter who was standing in front of his eyes, but he could see the serial number of the watch which was hidden from his eyes by his daughter's body. So what is real in this world? Reality is usually anything that is seen to be solid and anything that is broadly agreed upon.

This illusion our minds create for us is a prison for humans, restricting them to their personal desires and to affection for only the few beings nearest them. Our task must be to free them from this prison by widening their circle of compassion to embrace all living beings and all of nature. Homo sapiens hopes, dreams, illusions and goals for the future make life worth living for him. He won't give up his illusions easily. His illusion as a southern slave owner, Catholic priest, Communist, Capitalist, prisoner, disabled person, victim, body, are usually more valuable than his desire to know the truth. Because, if he knew the truth about himself, all his illusions would vanish and he would be left alone in a supreme state of serenity without a game. The game must continue! He won't give up his selfish loyalties to himself, his family, his religion, his government, his skin color, his language, and culture without a fight. He will literally perish in his struggle to maintain the divisions that keep the human race from being united and free from fear, hatred, and ignorance.

Our personal futures depend on keeping going and improving conditions. If we make it we'll regain our true selves and go free. If we don't, we are in for an eternity of pain and amnesia for ourselves and for our families, friends and the planet. We won't get killed; we'll get slaughtered over and over for ever and ever.

47) THE MECHANICS OF COSMIC BEINGS?

In each incarnation we nominally get to ride the earth about 75 turns around the sun before we return to our usual lives as spirits. The spirit world is our natural habitat, not the Earth. The big secret is this: we are not our bodies! Humans get very confused about this. We aren't even orbs or our pictures, which we carry with us from one lifetime to the next. Until now no one has known who we really are. We are cosmic beings, life-statics. A life-static is what you were before you mocked yourself up as a robot, or doll, or as an animal body with an orb. A life-static is just a dimensionless viewpoint. It is a total stillness with absolutely no form, matter, energy, space or time. But it can generate these things and make a copy of all of its creations. The collection of copies, or orb, that a being takes with him is called a reactive mind. When the life-static starts keeping lots of pictures and recordings close to itself it assumes the form of a sphere, or orb. That's what we see in the photos. The orb sphere is in the immediate vicinity of the viewpoint of the cosmic being. The being sees every direction at once. They are not like humans who peer out of two holes in their heads. Most of the time these disembodied beings are "out of sight out of mind." We don't put any attention on them or wonder what they are or what they are doing until they show up in our photographs. When I see the wars, crime, drugs, psychiatry, insanity and man's inhumanity to man, I wonder what the custodians are doing to contribute to mankind's woes. But these unseen beings have been with us, watching, working throughout history long before man came to this planet.

This is Dannion Brinkley with a woman who has an orb over her head. Dannion maintains that each of us are great, powerful and mighty spiritual beings with dignity, direction and purpose. He said, "I have been to the other side of the veil, and I know there is no such thing as death. Therefore, what becomes most important, to all of us, is how passionately and powerfully we live!" (www.dannion.com)

People want to understand God. But they think that God is ineffable, and could never be explained. That's because all they have been given is inhibition, especially by their orbs. Humans deny themselves; they conceive themselves as the little "I am nots." They think they are not powerful, not wise, not just, not noble and not able to create. Their orbs give them this false impression. If you don't make yourself equal to god, you can't perceive god; for like is known by like. Separate yourself from everything that is physical, and grow as vast as that immeasurable vastness; step beyond all time and become eternal; then you will perceive god. Realize that nothing is impossible for you; recognize that you, too, are immortal and that you can duplicate all things in your mind; find your home in the heart of every living creature; make yourself higher than all heights and lower than all depths; bring all opposites inside yourself and reconcile them; understand that you are everywhere; on the land, in the sea, in the sky; realize that you haven't yet been begotten, that you are still in the womb, that you are young, that you are old, that you are dead, that you are in the world beyond the grave; in all times and places, in all substances and qualities and magnitudes; then you will perceive god. You have to be out of the box to perceive god.

People get excited when they see a picture of an orb and they say it is unbelievably strange. But I'll tell you what is strange. They are looking at a representation of what they have created and they don't recognize it. That is what is so strange! Their reality has been so consistently and methodically programmed to prove to them that they are animal bodies that they haven't a clue! That's what I call high strangeness.

After all the trillions of years that they've kicked around in the Physical Universe, with the googolplex (trillions of trillions) of pictures that they took along the way to remind themselves of their adventures, with all the trouble that they took to cart those pictures around all that time, they now see a representation of their creation and they're dumbfounded. In spite of their determination to remember their friends, enemies and everything they came across, some super psychiatrists in space have drugged, beaten, electro-shocked and painfully hypnotized them into total amnesia! My god! What a total invalidation of a BEING!

Now they don't even know their own name or who they were before they stole their present body. That's impressive technology! We could really use that on Bin Laden. We could wipe out his identity and make him a really expert window washer! But let's get back to my point. I'm really impressed with the technology, the insidious use of overwhelming force and the resolute evil that these beings used to bring us down to our present mental condition of total amnesia, utter weakness and blind slavery. Of course, they took advantage of our laziness and irresponsibility. Without our help, we couldn't have been turned into slaves.

In their present state, people can see photographic images of their past friends, who are still free beings as disembodied spirits, and they aren't even reminded of their lives with them. Their former spirit friends, carrying orbs, have not yet decayed into meat body "containers." They are aware, free, they can go any place, see everything, and know everything. But the humans are trapped in their own minds, trapped in their own heads, trapped on the surface of the Earth, trapped in this solar system, trapped in this Milky Way Galaxy and trapped in the Physical Universe. Their own universe has been brought to nothing and the Physical Universe has totally overwhelmed them with its all-consuming force, space, mass and time.

Imagine what life would be like without a body. Maybe you can recall a time between lives where you could see every direction at once, you could go anywhere you wanted just by thinking of a location, you could light up like a fire-fly if you desired, and you took all your lives experiences with you in the form of trillions of years worth of mental image pictures complete with sound, pain, taste and touch.

I am a cosmic being. When I was very young I used to dream of flying through the air over everyone's head because I had just come from an existence where I had done just that. Everyone has a mechanism for recording their life experiences and then playing them back when they die. You hear about it from people who have had near death experiences. That recording is placed in the orb's archives along with the other trillions of years' worth of recordings that the being has forgotten how to replay.

There are techniques available to clear away a human's own barriers to spiritual perception but it takes a very responsible person to utilize them. The technology is available today to make every human being totally free. I was freed from a debilitating hypnotic state on May 10, 2005 when I saw that the whole agonizing future of our planet depends on our taking responsibility for our fellows. A huge mass of my orb dissipated while I was overcome by grief and mental charge. It blew a big chuck out of my orb. When the charge cleared I was a new person. I was set free, released. I reversed a two hundred trillion year dwindling spiral that turned me into a powerless idiot with an insane attitude. I changed from not-knowing to knowing in a flash. Now I can see things as they are and there are no more mysteries for me. Those who are looking for freedom

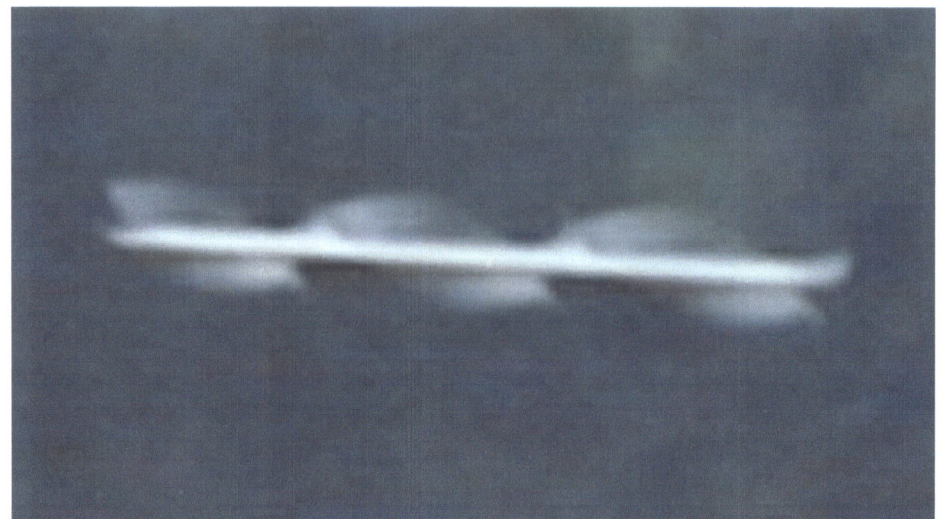

and knowledge will find it. You just have to keep your integrity intact and put one foot in front of the other until you arrive. You'll have to learn to flip the switch that powers up that 500,000 megawatt power generator – you. The picture (left) and the one below show images of rods. Rods are the 200 mph entities that have recently been found and photographed that are so light that they can fly right through the Earth. They are also called Sky Swimmers and they resemble squid or small bugs. They move faster than the eye can see but not beyond the perception of camcorders. They are one of a million species that live and move about us in the atmosphere and in space which are unseen by humans.

We are all cosmic beings. We've all committed atrocities in our distant past. It was the regret for our crimes that drove us to the degradation of becoming Homo sapiens. None of us could face our past, so trailing clouds of glory we settled for being human.

The game we play is "let's pretend", but we pretend we're not pretending. We decide to forget who we are but then we forget that we've forgotten. We are the viewpoint that watches and evaluates and directs the show. We are the star in our own production and there are no extras. We are the "I AM" consciousness. It's that powerful, loving, perfect reflection of the cosmos. But in our attempt to cope with early situations we chose or were hypnotized into a passive position. We put ourselves down, and have learned to tolerate this masochistic posture of weakness and indecisiveness. We completely denied ourselves. We denied all our abilities to produce our own material, energy, space and time and took a body that used the physical universe's material, energy, space and time. We figured that any game was better than no game. But in reality we are free, a center of cosmic energy. Our will is our power but we pretend we don't have it. So now we are lost out here in the nether regions of the Milky Way Galaxy on this prison planet. Sounds crazy doesn't it? But every orb researchers probed for this data revealed a history of space travel, planet building, terra-forming and fantastic technology. If so many people agree on this, it's real, even if most of us have had our memories tampered with and we cannot recall it consciously.

Our past lives have been expunged from our memories and in its place other peoples' memories have been imprinted in our minds. We are eternal, spiritual beings experiencing the human condition, so we assume that we are human bodies and we chose for ourselves human goals, not divine goals. We create a cute little house in the suburbs with a 30 year mortgage instead of devising a plan to get us and our friends out of this insanity. We have no idea who we are because our minds are so muddled.

This knowledge isn't pretty. The truth of our past is very difficult to confront. You could get a headache or some other bizarre pain, get sick and throw up if someone told you what you've been through. We've done a lot of things that we sorely regret. And we've turned, at times, in the eons past, to destructive technologies to help us forget. We don't remember when we were in a better, more powerful condition. Reptilian beings don't want us to remember. The pain, drugs, hypnosis, cunning and shear force of energy used to knock us out was overwhelming. It takes a tremendous amount of energy to knock out a really big being, and no one has ever allowed you to know how wonderful and great you really are. You've no idea how much you have been inhibited. You just have no inkling of your potentials and your abilities, or the degree of oppression you've endured.

Psychiatrists would have you believe that most of human behavior is completely determined by the first few years of a child's life. It is true that early inadequacies in nutrition or intellectual, emotional, or physical damage can affect development, but all the fears and hang-ups that we blame on our treatment during childhood are open for restatement, redefinition, and remodeling by our belief systems. We are not the pawns of our upbringing if we don't want to be. We are actually free to be, do, and have anything we want as long as we believe it is possible and are willing to put in the effort and discipline necessary to bring it about,

That's not all; Homo sapiens see life and reality through the limiting perspective of a microscope.

COSMIC MECHANICS

They make mountains out of mole hills while ignoring the unifying, harmonizing macrocosmic realities that lie just beyond their limited views. It's like they're wearing blinders. Blinders are used on horses to keep them from being frightened by what they would see if they had broader vision. People do the same thing. They keep their blinders pulled in close enough to block out or condemn things that are different from what they are used to. This leaves them with a very limited but comfortable microscopic view of reality instead of a limitless and challenging

macroscopic view of themselves, others, and their relationship to the universe. You are actually as big as the number of viewpoints that you can have until ultimately, you are like God, who experiences all viewpoints.

(Photo above) "1995-- My boyfriend & I had several poodles while living together. Brianna loved to jump up & get her nightly "scratchdown", then when the heat completely melted her, down she'd jump & curl up for a nap. I knew he had these abilities, but never dreamed I'd catch it on film! The first photo showed nothing, then when she started to get that "goony" look on her, we laughed & I said I better get another one. To my amazement, I saw the energy actually coming down from the ceiling, over his shoulder & out the solar plexus to her. It's one of my treasured photos...."Jane B.

48) YOUR MOST VALUABLE POSSESSION IS YOUR UNDERSTANDING

More than anything, I value understanding. As I see it, the only richness that exists in the world is understanding. When you've got that, you've got everything, health, wealth, leadership and the

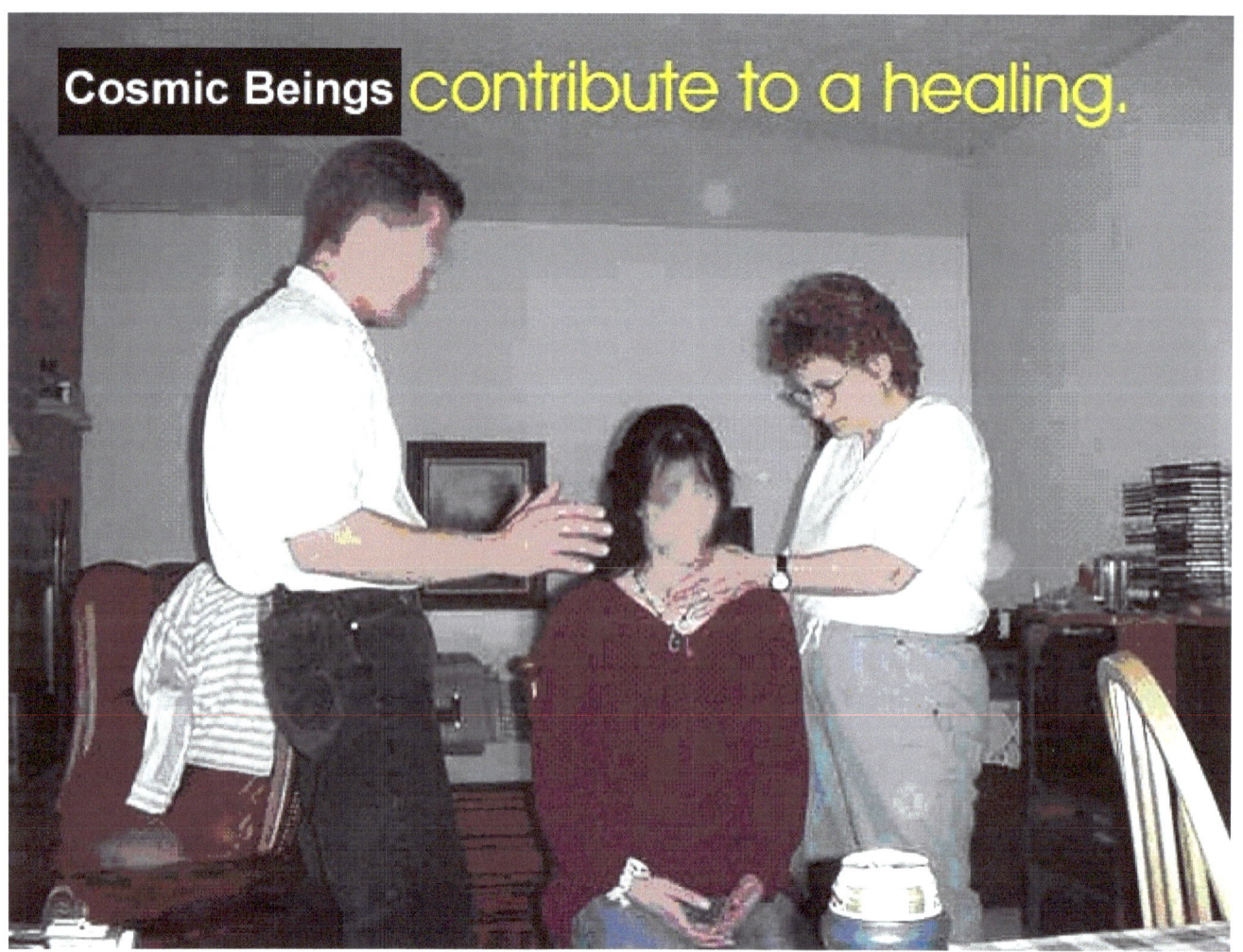

best of games. The only healing agency in the universe is the spirit. Above, is a picture of several cosmic beings with their orbs contributing to a healing.

Don't believe these pictures just because they are in print. We must be skeptical and check things out. The Billie Mier UFO hoax, Doug and Dave's hoaxed crop circles, hoaxed ghost photos and millions of other con jobs including landing Americans on the Moon during the Apollo Program has disabused me of my faith in mankind. Everything has to be checked out. Trust no one.

It seems like everything we know to be true is wrong. The "Authorities" tell you that pasteurized milk is better, cooking doesn't hurt the nutritional value of food, psychiatry improves mental health, the United States is a democracy, we pay taxes to support our government and protect our people, only the authorized experts should attempt to repair the mind, mankind couldn't get along without drugs, we're winning the war on drugs and a larger police force will keep crime down. This is craziness, as is the discrete particle of mass theory.

49) THE MECHANICS OF ENERGY PICTURES

Energy pictures are attached to ridges that you can see in the digital photos of 21 gram orbs. The pictures are actually interference patterns similar to holographic emulsions of quantum scalar waves that cosmic beings can see when they scan energy over them. Scanning the orb with light does not produce an image. The ridges in the orb must be scanned with much shorter wavelengths of live zero-point energy to produce the pictures. You do the same thing when you recall your last birthday party. You get the picture in your mind by scanning the interference pattern you made at the time of the party. The pictures go on back some 200 trillion years. The standing waves which comprise the ridges actually have a very small amount of mass. Everything that exists in the Physical Universe exhibits the phenomena of mass or inertia and the orbs definitely have physical mass. Particles range in density or weight from electrons to mental image pictures, to orbs, to rods, atoms, molecules, animals, rocks, planets, suns, galaxies etc.

Human's make mental image pictures at a rate of about 25 per second. The cosmic beings love their pictures the way we love our TV and cinema and they take them along everywhere they go. They collect pictures much the same way that we collects stamps, LPs, CDs, or videos.

Cosmic beings can rip chains of pictures away from other being's orbs. This is outright theft and it damages the victim. Modern examples of artistic theft are the MP-3 downloading of all the artists on the Web. This is how humans steal the valuable contributions of our greatest artists. Software, brand names, movies, books and inventions are similarly stolen. Imagine how it would feel if a powerful being had a tractor beam on you and your orb was ruptured. You were losing billions of your cherished pictures per microsecond as the beam jerked them away and your only hope was to cut your losses and run. Wounded and hurt, you decided never to go into that galaxy again. Now you have reduced yourself and your reach. You continue to get smaller each time you have a run-in with beings in the Physical Universe because you decide that there are more and more planets that you cannot BE. Finally you paint yourself into a corner and decide that you can no longer Be a part of the material universe (space), can no longer use force (energy) because whenever you use it someone gets hurt, can no longer be anything (mass and time) because it always winds up smashed. So, what do you do? You decide on a substitute beingness, an alter-ego. You pretend to be a body with an orb. The only energy you have is the energy of the body's engine, the only space you have is the space taken up by the body, the only matter you have is the mass of the body and the only time you have is the time dictated by the frequency of the electron.

Man, you are powerless and stuck! What happened to that 500,000 megawatt power generator you used to be? Look how far you've come from a cosmic being that can put suns in their orbits around galaxies. Look at the Hubble deep space images to see what other cosmic beings have produced. The Ring Nebula, 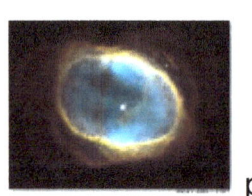 pulsars,  magellanic clouds,  the Horse head Nebula,  colliding galaxies  and supernovas are on the list of creations by big beings. The beings that you catch with your digital camera are degraded. Over trillions of years they have pulled in their boundaries until now they are only inches across and have lost their energy. It would be impossible to photograph a truly powerful being. It could be bigger than our solar system or even bigger than the whole Physical Universe, but you could photograph its creations.

When a cosmic being gets a new body it fastens itself to the motor controls on the left and right sides of the brain's cerebral cortex. Once attached to a human, a hypnotist can call up any number of mental pictures that the being has stored away over the last trillions of years. Orbs help to explain the phenomenon of people who suddenly, out of no observable training, begin to speak foreign tongues, men who "seem to remember having been here before," strange yearnings in people for various parts of the country or the world or the stars of which they have no actual knowledge. Many times feelings of loneliness and isolation are only the separation we feel from our dear friends in the spirit world. It is therapeutic to help a person recall his prior lives and friends, but it is far more meaningful to help them find their place in the spirit world of zero-point energy, their true home. The ultimate therapy is to give a being total freedom by getting rid of his orb. This can only be done with live communication from another being. You cannot do it yourself because over 200 trillion years the orb has grown more powerful than you. It won't go away without a fight. You'll lose. It is set up that way. You never had a chance out of the trap until now. You'll need help. Contact the author to find groups in your area.

The brain is not the center of intelligence of Homo sapiens and it is not the repository of memory. Actually, each cell of the human body is a separate living animal all by itself and it has a mind of its own. It also has a brain in the cell membrane. The minds of all the cells of the body are connected by biophotonic communication via the various command centers to you, a cosmic being, who is supposed to be in control. You are the "I" of the body, but it is not you who are in control. The orb is in control and you are lucky if you have ten original thoughts of your own in a week! The orb runs everything by automatic circuits! You are the Captain of that ship, that low temperature carbon/oxygen engine and chassis that you call your body. But you are not in control. That is why people who get heart transplants take on the attitudes and tastes of the donor. You may have heard of the woman who took up beer drinking and skiing in Switzerland after she had a heart transplant from a Swiss beer-drinking skier. The male donor was anonymous but after a couple of years his identity was revealed and the lady discovered the source of her strange yearnings and activities.

One conservative, health conscious woman from New England by the name of Claire Sylvia was astonished when she developed a taste for beer, chicken nuggets and motorcycles after her heart-lung transplant. Sylvia found out from the donors family that she had the heart of an eighteen-year-old motorcycle enthusiast who loved chicken nuggets and beer. Sylvia outlined her personal transformational experiences and the experiences of other heart transplant patients in her book, A *Change of Heart*.[Sylvia and Novak 1997] There are a number of other stories in *The Heart's code: Tapping the Wisdom and Power of Our Heart Energy*. [Pearsall 1998] These transplanted memories are exceedingly accurate and beyond chance or coincidence. In fact, after her heart transplant, one young girl began having nightmares of murder. Her dreams were so vivid that they led to the capture of the murderer who killed her donor.

Cells are physically endowed through their zero-point entities with perception mechanisms that can distinguish and remember a taste for chicken nuggets. Your cells can acquire a taste for Coke or Drugs or Beer, and put a demand on you to feed it that product. You think that you are addicted to a drug, but in fact it isn't you, but your cells that are hooked on it. If you don't put your cells in line and discipline them, they will mutiny and leave you homeless and unconscious as a drunk in the gutter. You are the captain of that ship, it is your job to keep those cells in line. If you don't, you will be stuck with toothaches, backaches, painful joints, lethargy and poor health. Each cell has wonderful biophotonic communication with the other cells so it can coordinate in achieving survival for the organism. The biology of life is awe inspiring.

Wouldn't it be wonderful for all of us to discover the source of all our strange and aberrated behavior? How can we tell which cell or control center or past life personality of the orb is making the decisions? If we could work that out and discover our own true identity, our thinking would be cleared up forever and our amnesia would be gone.

50) WATER HOLDS A MEMORY

Water is paramount on Earth, if not in the universe. We live on a water planet. Our bodies are mostly water. The water molecule has an angle of 105 degrees, which is in the golden mean proportion. Unstructured water is missing one electron from its outer orbit, whereas structured water has no missing electrons. A healthy frozen water crystal is shown at the left. In the human body, there are two basic types of water (biowater) - Bound and Clustered. This is a frozen crystal of clustered water.

When we are born, the trillions of young cells in our bodies are filled with a special kind of water. This special water is "clustered" in beautiful, star-shaped molecules.

The perfect shape of these water molecules allows them to slip easily through the cell membranes, carrying nutrients into the cells and carrying waste and toxins out. This easy passage of star-shaped water molecules in and out of our cells is what keeps the cells clean and young and vibrantly healthy.

The DNA in each of our cells, which determines how we grow and what we look like, is folded around a core of this remarkable water.

As we get older, the star-shape of the water molecules changes. Because of stress and pollution, toxins and accumulated wastes in our bodies gradually bind to the water molecules, enlarging and distorting them. As the normal clustered water molecules in our bodies get bigger and lose

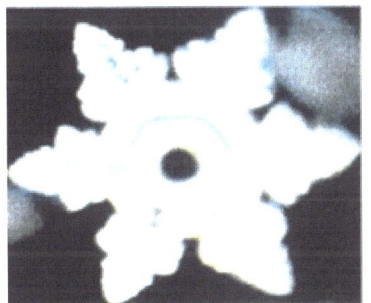

shape, it becomes difficult for water to pass the cell wall barrier. The cells of our bodies, once highly lubricated and cleansed with normal clustered water molecules, slowly gather more and more toxins and waste. This is a frozen crystal molecule of the purest water.

Some people drink clustered water as it supposedly hydrates, detoxifies, increases energy and mental clarity, promotes a sense of well-being, increases metabolism, feels great, helps lose weight and softens skin and hair.

Naturalist and scientist Johanne Grander revealed that water is like a liquid tape recorder storing information and vibrational frequencies, even after the physical substances or pollutants had been removed.

Imagine if water could absorb feelings and emotions or be transformed by thoughts. Imagine if we could photograph the structure of water at the moment of freezing and from the image "read" a message about the water that is relevant to our own health and well-being on the planet. Imagine if we could show the direct consequences of destructive thoughts or, alternately, the thoughts of love and appreciation. *The Hidden Messages in Water* introduces readers to the revolutionary work of Japanese scientist Masaru Emoto, who discovered that molecules of water are affected by thoughts, words, and feelings. THE MESSAGES FROM WATER by Masaru Emoto describes his experiments with frozen water crystals taken from both natural and chemically-treated or polluted water sources, showing the crystalline patterns in frozen water give a message as to the water's basic structure and symmetry. Clean and pure water, from natural and unpolluted sources, show brilliant and sparkling snowflake patterns, rich in color and luminosity. Again, Emoto's data shows that the Physical Universe is more thought than things. As J. Z. Knight says, "Thought matters."

Love for Humanity

Amazing Grace   *Love of Husband and Wife*   Emoto's research makes it crystal clear that water indeed is alive and responds to our human vibrational energy, thoughts, words, ideas and music. His photos show the molecular structural changes in the water following various inputs/impressions. He freezes droplets of water and then examines them under a dark field microscope and takes photos. His work clearly illustrates the effect of the environment upon the structure of the water. Here are a couple of examples of this:

"You Make Me Sick. I Will Kill You."
results in this structure in the sample of water.

1972 – The *Secret Life Of Plants* -- a fascinating account of the physical, emotional and spiritual relations between plants and man. It describes their ability to adapt to human wishes, to respond to music, their curative powers and their ability to communicate with man.

Dr. F. Lee Aeilts

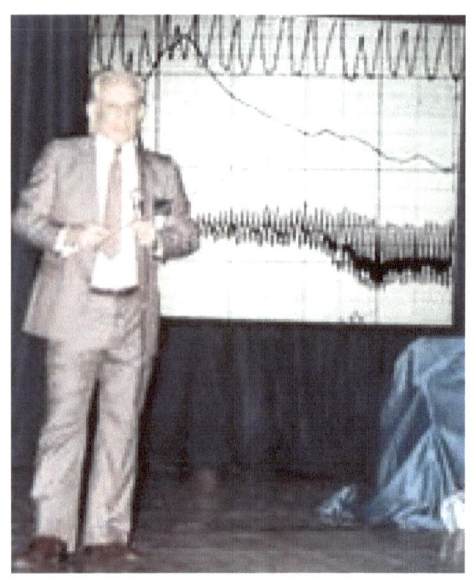

CLEVE BACKSTER 1969 (left) *"My plant read my mind!"* the FBI and CIA's then foremost polygraph researcher, reintroduced modern science to the sentient nature of our universe. With straightforward electronics that a student or garage-level scientist can replicate, he proved to humans that their thoughts and emotions affect the behavior of their own and other living cells.

51) DNA PHANTOM EFFECT

In the photo below, Tyler took his own picture and an image of a gray haired woman showed up in the frame. " *May 13, 2006 9:47P.M.--* I took a picture of myself after a boring game of the Ouija board in the downstairs rec. room. When I uploaded the picture from my digital camera, a "person" appeared in the picture. -- Tyler"

COSMIC MECHANICS

Between 1993 and 2000 a series of unprecedented experiments demonstrated the existence of an underlying field of energy that bathes the universe. Quantum biologist Vladimir Poponin reported the research that he and Peter Gariaev were doing at the Russian Academy of Sciences. They designed their pioneering experiment to test the behavior of DNA on photons. They first removed all the air from a specially designed tube; creating a vacuum.

Using precisely engineered equipment that could detect the photons, the scientists measured their location within the tube. They wanted to see if the photons were randomly scattered within the tube. They found the photons to be distributed in a way that was completely unordered. They expected as much.

Then they put samples of human DNA inside the closed tube with the photons. The particles arranged themselves differently in the presence of the live DNA. The DNA had a direct influence on the photons which wasn't predicted by conventional physics. When the DNA was removed from the cylinder the photons remained ordered just as if the DNA was still in the tube. Poponin described the light as behaving "surprisingly and counter-intuitively."

Poponin's new field structure reminds us of Max Plank's "matrix" identified 50 years earlier as well as the effects suggested in ancient tradition. "All matter originates and exists only by virtue of a force….We must assume behind this force the existence of a conscious and intelligent mind. This mind is the matrix of all matter." (Max Plank, 1944) Now we know that cells/DNA influence matter through this type of energy; biophotons. This is also the energy that cosmic beings and orbs use to perceive the sensations of the body. It comes from the thought universe, a universe of waves, not masses.

Doctors Iona and Alan Miller did more experimental work on the Phantom DNA Effect in 2002. They determined that a unified subliminal field of potentially universal consciousness apparently exists and may be explained as emerging from a previously overlooked physical vacuum or energy matrix. Currently, researchers have found no other substance which recreates or emulates the effect of the DNA molecule.

52) SPOOKY ACTION AT A DISTANCE

During the 1990s, scientists working with the U.S. Army investigated whether or not the power of our feelings continues to have an effect on living cells, specifically DNA, once those cells are no longer part of the body. They wanted to know whether thought, emotion or effort still impact tissue samples removed from the body. In a 1993 study reported in the journal *Advances*, the Army performed experiments designed by Dr. Cleve Backster, (above) to determine precisely whether the emotion/DNA connection continues following a separation. The researchers collected a swab of tissue and DNA from the inside of a volunteer's mouth. This sample was placed in a specially designed chamber that could measure the DNA electrically to see if it responded to the emotions of the person that it came from. The donor was several hundred feet away in another room.

The donor was shown a series of video images purposefully to create genuine emotional states within the body. The images ranged from wartime casualties to erotic and comedy footage. While the donor was experiencing these graphic images, the DNA in another room was being measured for its response. When the donor experienced "peaks" and "dips," his cells and DNA showed a powerful electrical response at the same instant of time. Whether the cells were in the same room or separated by hundred's of miles, the results were the same.

53) INTENTIONS CHANGE DNA

In 1991, an organization named The Institute of HeartMath was formed for the specific purpose of exploring the power that human feelings have over the body. Between 1992 and 1995 the researchers isolated DNA in a glass beaker and then exposed it to a powerful form of feeling known as coherent emotion. According to Glen Rein and Rollin McCraty, the principle researchers, this physiological state may be created intentionally by "using specifically designed mental and emotional self-management techniques which involve intentionally quieting the mind, shifting one's awareness to the heart area and focusing on positive emotions." The results were astounding. Human emotion changed the shape of the DNA! Without physically touching it or doing anything except producing precise feelings, the DNA responded by winding up or unwinding.

The implications of this experiment are mind boggling. We've been told that we've been born with a DNA set that determines our intelligence and physical attributes; that we are stuck with it. But this experiment shows us that nothing could be further from the truth. We can get smarter and tougher and stronger according to our intentions and desires. These experiments teach us that there is a zero-point field that connects any one thing with everything else in the universe. They also show that the DNA in our bodies gives us access to the energy and information that connects our universe. The zero-point field is made of an energy form that's unlike any we've known about in the past. Some call it "subtle energy." It doesn't work the way a conventional electromagnetic field does. It appears to be a tightly woven web and it makes up the fabric of creation, the zero-point field.

This is what we know of the zero-point field: 1) It is the cosmic source of the Physical Universe. 2) It is the bridge between our inner and outer worlds, the living essence that is the fabric of our reality. 3) It is the mirror that reflects our thoughts, feelings, emotions and beliefs. 4) It is everywhere all the time. 5) It has intelligence and consciousness. It did a wonderful job of creating all the fauna and flora on the Earth, including our bodies. 6) It is intimately connected to our health. 7) It's what space itself is made of. And what creates the zero-point field, you ask? We do; all the cosmic beings in the Physical Universe!

54) THE MECHANICS OF DNA AND RNA UNRAVELED

DNA is composed of regularly repeating subunits called nucleotides. Only a limited number of nucleotides exist in nature and all contain three elements: (i) a phosphate group(s) linked to a (ii) sugar, which was joined to a (iii) flat ring molecule commonly called a base. The limited number of natural nucleotides is partially a result of the fact that only five types of natural bases exist: guanine (G), adenine (A), cytosine (C), thymine (T), and uracil (U). Each nucleotide possesses the ability to link to others to form chains. Surprisingly, only two similar types of chains existed, DNA and RNA. The most obvious difference between the two types was that the base uracil was only found in RNA, while the base thymine was found only in DNA.

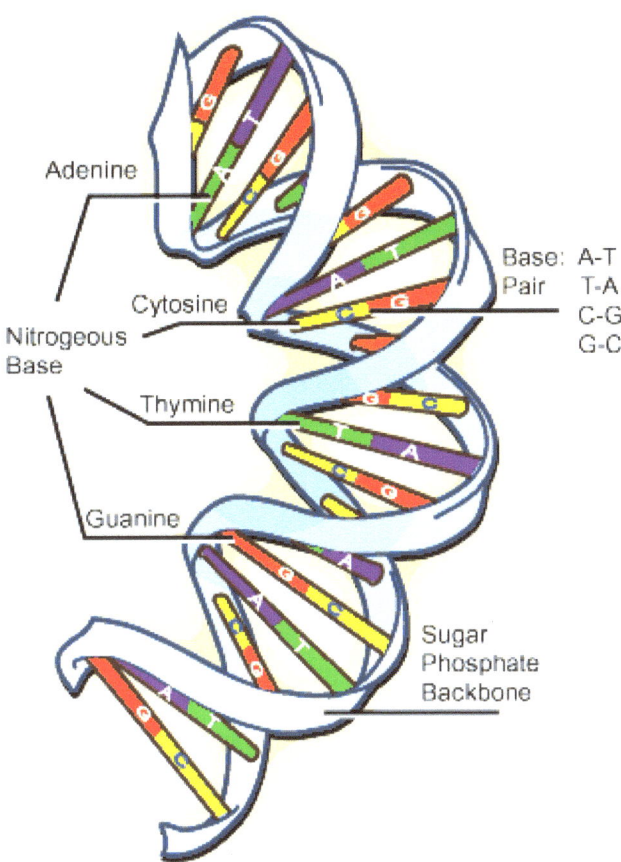

The DNA double helix.

How does DNA govern and dictate the natural variety of life on Earth? How does the chemical structure of DNA, essentially a chain of nucleotides linked together, enable it to act as a carrier of inheritance?

The bases of the nucleotides (G, A, C, T, U) can chemically bind to each other. Not only that, but they do so in an exceptionally specific manner. Adenine binds only to thymine in DNA (and uracil in RNA), while guanine binds only cytosine. As a consequence of this the amount of adenine equaled the amount of thymine, while the amount of guanine equaled the amount of cytosine in a DNA molecule. This was later known as Chargaff's rule, after the Austrian chemist Erwin Chargaff.

X-ray diffraction studies show that DNA adopted a regular and precise helical structure. Enough data was now in place for a famous leap of scientific faith to be taken. In 1953, Watson and Crick correctly deduced that DNA forms a double helix with two strands of nucleotides wrapped around each other (see above). The binding rules for nucleotides ensured that each strand was a complementary copy of the other (for example an adenine in one strand was always bound to thymine in the other strand). Thus, the two strands were complementary anti-parallel chains of nucleotides wound around each other to form a double helix. Such a structure provides DNA with a simple mechanism to accurately reproduce itself: just pull the two strands apart and use one strand to create a complementary copy of the other using the nucleotide binding rules. If done for both strands, two exact copies of the original DNA molecule are created. The realization that DNA formed a double helix solves a large part of the question of how DNA was involved in all life forms by revealing how to make endless copies of a DNA molecule

The Genetic Code

There is an important class of proteins, termed enzymes. These proteins are able to catalyze biochemical reactions and are found to be responsible for biochemical function. Proteins are composed of 20 naturally occurring amino acids linked together in a chain (called a polypeptide), much like DNA.

Studies of defects in well-known metabolic reactions show that single genes direct the production of single proteins, a fact which is now generalized as the "one gene = one protein" rule. This rule and the understanding that genetic information is specified in the four nucleotide bases of DNA (A, C, G and T) causes one to wonder how four bases of DNA could encode the 20 known amino acids that make up proteins? The linear sequence of nucleotides in a DNA strand corresponds to the linear sequence of amino acids in a protein polypeptide. This is evident because the position of mutations in a protein correlate to the position of mutations in a gene (i.e. they appear in the same relative places in the molecules). A co-linear relationship exists between the two. In 1961, a historic set of experiments were begun by Marshall Nirenberg and Heinrich Matthaei that marked the beginning of modern molecular biological techniques. They were able to create a synthetic nucleotide chain composed only of uracil (i.e. UUUUU), which they went on to add to cells that had been broken apart. When they did this they witnessed the production of a polypeptide chain. Even more interesting, it was composed of only a single amino acid, phenylalanine. Next, they began adding defined lengths of uracil chains to the extracts. By doing this they found that only multiplies of three nucleotides produced amino acid chains. For example, a chain of three uracils (UUU) gave a single phenylalanine, similarly a chain of four or five uracils (UUU-U or UUU-UU) also produced only one phenylalanine, while a chain of six (UUU-UUU) produced two phenylalanines linked together. Hence it was determined that a triplet of uracils in a gene coded for the amino acid phenylalanine in a protein. Production of all the possible combinations of three nucleotides (called a codon) soon revealed which triplets coded for which amino acids. It was found that 61 combinations coded for the 20 amino acids, while the three remaining codons were used as "stop" signals for the end of a protein.

The genetic code was broken. Scientists understood how a protein is encoded in the molecular structure of DNA. With this information the underlying mechanisms of how a cell used DNA to make protein became clear.

Producing Genetic Messages from DNA

All living organisms depend upon the production of proteins encoded by the information held within their DNA. Despite the variations that exist between organisms, all cells make use of the same general mechanism for decoding the information in DNA into proteins, termed gene expression. Even though the amino acid sequence of proteins is dictated by the nucleotide sequence of many genes, proteins are not directly synthesized from DNA. Instead, genes produce proteins in two discreet stages, which involve many different types of enzymes, proteins and RNA molecules.

The first stage is called transcription (see below), in which an RNA copy (or transcript) of a specific gene is produced. This RNA copy of the gene is called a messenger RNA (mRNA), since it is the genetic message that will produce a protein. Production of mRNA requires an enzyme called RNA polymerase. It begins the process by binding to specific nucleotide sequences in the DNA (called a promoter), located just up from the gene that specifies a protein. A complex process unwinds the DNA in this area so that the polymerase can begin to move along the DNA strand like a train along a rail. As the RNA polymerase moves along it synthesizes an RNA copy according to the nucleotide sequence it encounters (done by pairing a new nucleotide to a complementary nucleotide in the DNA using the base pairing rules, followed by linking them together). This procedure continues until the polymerase hits a defined sequence called a terminator, which causes the polymerase to fall off the gene and release the mRNA. Once released, an mRNA is free to float through the cell bearing its genetic message and ultimately engages in the second stage of producing a protein from a gene.

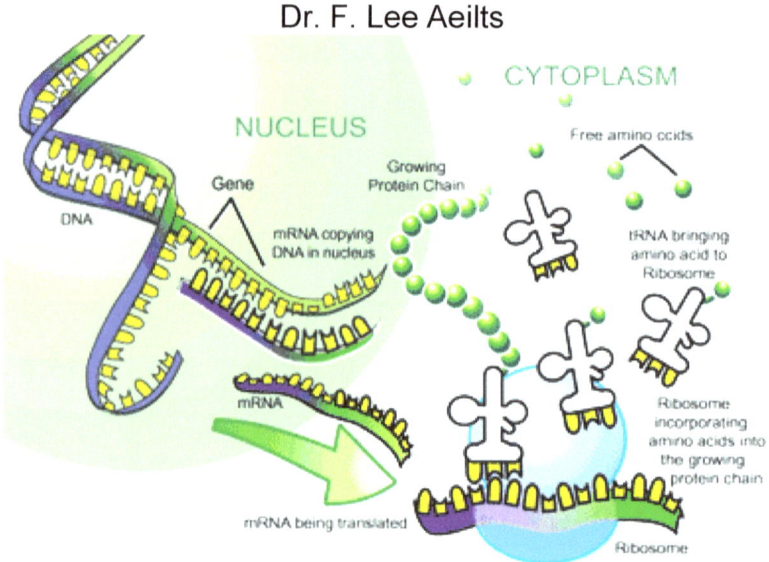

DNA transcription and translation.

Decoding Genetic Messages

The process of decoding mRNA transcripts is termed translation and once again involves many types of proteins and RNA molecules. In particular, translation requires two types of RNA termed ribosomal RNA (rRNA) and transfer RNA (tRNA). Ribosomal RNA is intimately involved in the synthesis of proteins through the interaction of various types of rRNA to form a complex called a ribosome. This is the cellular machine that creates proteins from mRNA. A ribosome forms a donut structure with the mRNA passing through its center; a specific site within the mRNA (called the ribosome binding site or RBS) then binds to the ribosome, causing the second type of RNA, the tRNA, to spring into action. tRNAs are universal adaptor molecules that carry amino acids and a complementary triplet of nucleotides called an anti-codon that recognizes each codon in an mRNA. As mentioned before, codon is the name given to triplets of nucleotides in mRNA that code for particular amino acids. An anti-codon is just a complement of a codon. Through this method each triplet in an mRNA molecule will bind to a tRNA that bears its complementary triplet codon. For example, a string of A-U-G nucleotides in an mRNA will bind a tRNA that has a U-A-C triplet, all of which is based on the nucleotide binding rules. Since each tRNA molecule carries an amino acid, the triplet codon will result in a specific amino acid being brought to the ribosome. At the ribosome these amino acids are bound together into a polypeptide chain (i.e. a protein) in exactly the linear order that the mRNA dictates, a sequential process that will produce a protein corresponding directly to the nucleotides in the mRNA.

COSMIC MECHANICS

Are you getting the feeling that there is a conscious creative intelligence directing the profusion of activities involved in the creation, monitoring and administration of your body? Can you feel admiration for the being that is doing this? There is a cause in the thought universe that is creating quantum waves to build wave structures and manage all life forms in the Physical Universe. Many people call that cause God, but God is poorly understood. If we could get rid of our orbs and return to knowingness, we would find that we cosmic beings are part of God.

That the living organism is coherent as a whole is not surprising. What is surprising is the degree and form of coherence. The organism's coherence goes beyond that of a biochemical system and in some resects it approaches the coherence of a quantum system. If an organism is to survive, its component parts and organs must be precisely yet flexibly correlated with each other. As long as it is living, it is in a state of dynamic equilibrium in which it stores energy and information and has them available to drive and direct its vital functions.

In a complex organism the challenge of order is gigantic. The human body consists of some 75 trillion cells, far more than stars in the Milky Way Galaxy. Of this cell population, 600 million are dying and the same number are regenerating every day. Each regenerated cell must inherit the mind of the previous cell to recall the ways of survival and avoid the pitfalls that lead to death. Over ten million cells per second are being constructed to aid the survival of the organism. The average skin cell lives only for about two weeks; bone cells are renewed every three months. Every ninety seconds millions of antibodies are synthesized, each from about twelve hundred amino acids, and every hour 200 million erythrocytes (red blood cells) are regenerated. The substances that coexist at a given time produce thousands of biochemical reactions in the body each and every second. Those cells are intelligent and diligent, miraculously working like crazy to keep that body going on the micro level. And what do we find on the macro level? The guy is a couch potato drinking beer and eating popcorn while watching *As The World Churns* on TV.

What will the Human genome project do?

Most Americans barely understand the social and political implications of "mapping" the human genome whereby geneticists can rewrite the vocabulary of life. DNA serves as the energetic antenna for the Life Force. The ignorant believe that genetic tests and therapies promise wholesale cures. There is no scientific evidence for this. The truth is that all genetic engineering and gene therapy is experimental and dangerous. A growing list of organic and environmental casualties speaks for this imperfect eugenics practice and guinea pig science. Heaven forbid that the secret societies on our planet steal the patents and copy rights on the genome discoveries and use them to oppress us all.

The idea that scientists can simply "map" the complex human genome and derive a medical utopia is ridiculous. Electro genetics and life is incredibly complex. Each of the 75 trillion cells in your body communicates with all the others simultaneously like a wireless, fiber optic phone network in four different languages. Are doctors going to manipulate this system? The Holy Grail of Biology is the human genome comprising three billion letters. If the genome were a story book it would consist of six billion bits of information in 23 Chromosomes or 23 chapters in your organic book of life. Each chapter has several thousand stories (30,000—40,000) called genes. Each story is composed of paragraphs, called exons, which are interrupted by subtitles called introns. Each paragraph is made up of words called codons. Each word is written in the letters A, T, G, and C. That's DNA's genetic alphabet called amino acid pairs. These base pairs pair up with one another (A pairs with T) (G pairs with C) to form the twisted ladder-like double helix. Thus, your body is unique.

There are about one billion words in your storybook, which is equivalent to 800 Bibles. It is astonishing that this genetic book of life fits inside the microscopic nucleus of a cell that fits easily on the head of a pin.

Biotech evangelists want you to believe they have a definite handle on this and can effectively manipulate this, without even diagnosing the electromagnetics and bioacoustics of the genome's origin. This is patently untrue. More important than your genetic code, or genotype, is your phenotype – the resulting physical expression of genetic activity that is mostly determined by exposures to other things including: stress, diet, cellular immunity, free-radical assaults on your DNA, behavioral factors such as beliefs, attitudes, thoughts and emotions, efforts and counter-efforts, toxic vocabulary, and physical factors including history of infections, injuries, inflammations, pH, dehydration, and toxicity levels. And don't think that sums it all up because I've neglected the two primary factors of your phenotype – you and your orb! You abuse your body and stress it out more than all the other factors except for the orb. The orb works 24/7 to do you in and keep you down. Your orb is the single source of all your problems, stress, unhappiness and self-doubt. It stores all your painful experiences and then uses those against you. Your DNA is severely stressed by your orb. Get rid of it!

Profound advances in the healing arts and sciences, both in diagnosis and treatment, are assured by quantum field technology. Sound and quantum fields alter genetic expression and water structuring so that DNA can broadcast what amounts to love songs from a universal orchestra. It doesn't matter whether you wish to acknowledge yourself as the Master Conductor of this symphony or not, you still get to enjoy the concert.

We now recognize the creationistic theories and spiritual healing practices previously belittled as "pseudo-science" and "quackery" are now firmly supported by hard science. The structuring and reshaping of matter through sound frequencies and energized quantum wave structures vibrate our life forms into physical realities. It only takes one step further to recognize yourself as a cosmic being, using zero-point scalar waves and quantum electromagnetic energies to materialize your form as a divine, although temporal, holographic apparition.

55) THE MECHANICS OF THE FIELD

Lynne McTaggart wrote *The FIELD, The Quest For The Secret Force Of The Universe*. (ISBN 0-06-093117-5) She shows that on our most fundamental level, the human mind and body are not distinct and separate from their environment but a packet of pulsating power constantly interacting with the vast energy sea. It helps us understand the visual human aura, the human memory and the power to heal. She describes experiments that prove what religion has long held to be true; that human beings are far more extraordinary than an assemblage of flesh and bones. "At our most elemental, we are not a chemical reaction, but an energetic charge. Human beings and all living things are a coalescence of energy in a field of energy connected to every other thing in the world. This pulsating energy field is the central engine of our being and our consciousness…"

Everything that exists has being, some divine essence, and some degree of consciousness. After all, it is built of quantum wave structures coming from the thought universe of creative intelligence and consciousness. Even a stone has rudimentary consciousness, or it would not exist. Something is maintaining its form. Existence is not an accident of space and time. Everything is alive, the sun, the earth, plants, animals and humans. All are expressions of consciousness to varying degrees—consciousness manifesting in form. **The world and everything that you**

experience is created by the interplay of the static with the dynamic. You are the divine static of perfect stillness; the dynamic is the zero-point field and world of motion around you.

How separate are you from the universe?

It depends on if you mean you, the cosmic being, or you your body. If a fish is born in your aquarium and you call it John, write out a birth certificate, tell him about his family history and the struggles of his ancestors, he becomes a special personality to you. You granted him beingness. If one minute later he gets eaten by a bigger fish your sense of fairness and justice is assaulted by the tragedy. You wonder how God could let such a thing happen. But that is God's way, and to the fish in the aquarium you are God. You create the aquarium and put all the fish in it. The death only appears tragic because you projected a separate self where there was none. You witnessed a tiny portion of the dynamic process, a molecular dance, and made a separate entity out of it.

Consciousness takes on the disguise of forms until they become so complex that it completely loses itself in them. In present day humans, consciousness is completely identified with its disguise. It only knows itself as form and lives in fear of the annihilation of its physical or psychological form. This is a manifestation of the orb limiting our awareness and it causes considerable dysfunction.

Although you are a cosmic being with a universe peculiar and unique to you, there is no 'me' and 'not me' duality to our bodies in relation to the Physical Universe, but one underlying energy field. Just as there is one underlying energy field in your own universe, the common universe has one underlying energy field. Your body isn't made of the stuff of your own universe. It is made of the stuff of the Physical Universe and built so that cosmic beings can control it with photons. When the body falls over, dead, the cosmic being leaves it to be recycled by the local fauna and flora, and finds another fresh body to continue its adventures in the material universe. **The field is the force, rather than germs or genes that finally determine whether we are healthy or ill.** It is the force which must be tapped in order to heal.

The scientific materialism of Newton and Descartes denied God and Life from the world of matter. They ripped the heart and soul out of the universe. Charles Darwin made matters worse with his theory of evolution. He presented life as random, predatory, purposeless and solitary. Be the best or die. We are an evolutionary accident. Eat or be eaten. We are genetic terrorists, efficiently disposing of any weaker links. He said life is about winning, getting there first. Get in, hit hard and get out. Man did not claw his way to the top of the evolutionary line to be a vegetarian.

Scientific materialism has denied us answers to the most fundamental mysteries of our own being; What is life, how does it begin, what is a mind, why do we get ill, how do we control our bodies, what happens to us when we die? A few quantum physicists picked up where the pioneers of quantum physics left off. They probed deeper. They examined the few equations that had been subtracted out in quantum physics. These were the equations of the zero-point field, an ocean of microscopic vibrations in the space between things. The zero-point field, the very underpinning of the Physical Universe, was a heaving sea of energy – one vast quantum field. This field connects everything with everything else like some invisible web.

Our bodies are made of the same basic material. They are composed of packets of quantum energy constantly exchanging information with the inexhaustible energy sea. Living cells emit a weak photonic radiation, and this is the most crucial aspect of biological processes. Cosmic beings could create photonic communication with the cells to control and heal bodies. Human perception occurred because of energy exchanges between subatomic wave structures of our cells and the quantum energy sea. We literally resonate with our world.

The coherence of the organism is quintessentially pluralistic and diverse at every level, from the tens of thousands of genes and hundreds of thousands of proteins and other macromolecules that make up a cell, to the many kinds of cells that constitute tissues and organs. There are no controlling and controlled parts or levels; all components are in instantaneous and continuous communication. Each cell knows its job. Each cell gets corrected or repaired when it fails or gets injured. If only the employees of a large corporation were so capable! As a result, the adjustments, responses and changes required for the maintenance of the organism propagate in all directions at the same time. This kind of instant, system-wide correlation cannot be produced solely by physical or even chemical interactions among molecules, genes, cells, and organs. Though some biochemical signaling—for example, of control genes—is remarkably efficient, the speed with which activating processes spread in the body, as well as the complexity of these processes, makes reliance on biochemistry alone insufficient. The conduction of signals through the nervous system, for example, cannot proceed faster than about twenty meters per second, and it cannot carry a large number of diverse signals at the same time. It has an extremely narrow bandwidth. Yet there is evidence that the entire organism is subtly but effectively interlinked; there are quasi-instant, nonlinear, heterogeneous, and multidimensional correlations among all its parts.

No matter how diverse the cells, organs and systems are, in essential aspects the act as one. That is coherence. According to Mae-Wan Ho they behave like a good jazz band, where every player responds immediately and spontaneously to however the others are improvising. The super jazz band of an organism never ceases to play in a lifetime, expressing the harmonies and melodies of the individual organism with a recurring rhythm and beat but with endless variations.

The system wide coherence of the organism provides further evidence for the quantum postulate. It is known that correlation can occur between distant molecules and molecular assemblies only when they resonate at the same or compatible frequencies. Whether the force that appears among such assemblies is attractive or repulsive depends on the given phase relations. For cohesion to occur among the assemblies they have to resonate in phase—the same wave function must apply to them.

When you lift your arm, stretch your leg or turn you head you are using zero-point energy from your spiritual viewpoint to control your muscles. That's why heavy electromagnetic fields in your vicinity don't disturb your coordination. You are not controlling with electromagnetic fields. They can fry your tissue, but they don't constitute a control signal.

SUPERCONDUCTIVITY

In the living organism, processes suggestive of superconductivity appear at macroscopic scales and normal temperatures. The detailed mechanism underlying this phenomenon is under intense research.

DNA in an Electrical Circuit

Your television is graphically animated because it receives energy signals through thin air.

Similarly, your liquid crystal components, within your cells and extra cellular matrix of your body, possess many of the features of televisions, computers and electronic circuits. Components analogous to conductors, semiconductors, resistors, transistors, capacitors, inductor coils, transducers, switches, generators and batteries exist in you and in all biological tissues.

Examples of components that allow your cells to function as a solid-state electronic device include: transducers (membrane receptors), inductors (membrane receptors and DNA), capacitors (cell and organelle membranes), resonators (membranes and DNA), tuning circuits (membrane-protein complexes), and semiconductors (liquid crystal protein polymers). You control these circuits with quantum waves. Electrical and magnetic frequencies, and sub sensory sound sourcing from what amounts to a "Universal Orchestra", moves mountains of physical matter along paths of least resistance flowing into geometric forms that you perceive as the Physical Universe.

The Electrical Roles of Membranes and Mitochondria

Electricity in your body comes from the food that you eat and the air you breathe. The process is called metabolism and works analogous to a car engine burning fuel to produce energy. Cells derive their energy from enzyme-catalyzed chemical reactions which involve the oxidation of fats, proteins, and carbohydrates. The oxidation (burning) of petroleum moves your car, the oxidation of fats, proteins and carbohydrates moves your body. Quantum wave structures found in the body include photons, electrons, protons, elementary ions, inorganic radicals, molecules and molecular aggregates. Photons act on electrons by raising their energy state. (This is called excitation.) Excited electrons can drop back to more stable energy levels by emitting photons of light energy. Electron excitations can lead to the formation of electrical bonds between molecules. These are the traditional bonds of classic chemistry. The breaking of such bonds can, by reverse process, lead to the excitation of electrons and the emission of photons. On a dark night pull some black electrical tape off a roll and notice the little line of light that glows where the tape is separating from the roll. The electrons are being excited by breaking the chemical bonds of the adhesive. In living systems the excitation of electrons by photons from the sun is transformed into high energy phosphate bonds by the process of photosynthesis. The release of the energy stored in these bonds is the fuel of life. Electrons are transferred between molecules in a downward cascade fashion to lower energy states. This action produces the electrical current that produces the motion we call life. This also powers the building of biological structures. This process is monitored by a thought universe entity working through the DNA. You are not monitoring your cellular processes and managing the life of the body. An entity, not a cosmic being, in the thought universe directs the activities of the body with zero-point waves. You are the captain of that ship (your body), but the entity actually controls the genes and takes care of housekeeping functions. All living tissue has an entity in the thought universe which directs the DNA and guides the processes of regulation, repair, and regeneration. The reversal of the process of photosynthesis is called *bioluminescence.* During this process, energy is transferred from a molecular bond to an excited electron. Thus a light energy photon is emitted. Each DNA molecule produces these photons to communicate with the rest of the body. The Cosmic Being and the orb communicate with the body utilizing zero-point energy. That's why digital photos show orbs in the vicinity of the body with no apparent connection. The connection does not manifest in the material universe the same way that the connection that magnets have with each other does not manifest. Magnetic and gravitational forces are generated by the Zero Point Field between space resonances.

Dr. Bruce Lipton is a cellular biologist who did ground-breaking work in the field of new biology. His experiments, and those of other leading edge scientists, have examined in great detail the mechanisms by which cells receive and process information. In his book, The *Biology of Belief*, (ISBN 0-9759914-7-7) he shows that genes and DNA do not control our biology: that instead, DNA is controlled by signals from *outside* the cell, including the energetic messages emanating from our positive and negative thoughts. This profoundly hopeful synthesis of the latest and best research in cell biology and quantum physics shows that our bodies can be changed as we retrain our thinking. He presents experiments that show that the DNA is not the brain of a cell, but rather its gonads. The membrane of the cell functions more like a brain with its receptors and biochemical pathways. But that doesn't mean that the membrane is the mind. There is a great difference between a brain and a mind. Our own reactive mind is a Zero Point Field created device and the cell's mind is a much simpler Zero Point Field device. Genes are physical memories of an organism's learned experiences built by its zero-point mind. Structure is determined by function so the cell's history of efforts and counter-efforts has resulted in its genes.

Human beings have a great capacity for sticking to false beliefs with great passion and tenacity and hyper-rational scientists are not immune. Our well-developed nervous systems headed by our big brains and orbs provide us with a more complicated awareness than single cells. A cosmic being can choose to perceive the environment in different ways, although like cells, its orb makes its awareness more reflexive.

We can change the character of our lives by changing our beliefs. Dr. Lipton found a science-based path that would take him from being a perennial "victim" to co-creator of his destiny. The belief that we are frail biological machines controlled by genes is giving way to an understanding that we are powerful creators in our environments. Don't believe that if you are born with a defective happiness gene, you are destined for an unhappy life.

56) OUR MINDS HAVE BEEN COMPROMISED

None of us suspect that our minds have been compromised. A mind is such an insidious trap. When a hypnotist tells you that his pen weighs a thousand pounds and that you cannot lift it up, this instantly becomes a part of your reality. When the hypnotist brings you out of your hypnotic trance, he asks you to hand him his pen. You reach for it but you cannot lift it. You say, "This pen feels like it weighs a thousand pounds and I can't budge it. What have you done to the pen?" Notice that the subject never suspects his own mind has been tampered with, he just assumes the Physical Universe has been altered. When the subject is told about the post-hypnotic suggestion the subject will again be able to lift the pen.

We are so irresponsible and naïve as beings. We may have been kicking around this universe for trillions of years without improving our condition. What are you going to do about your condition this lifetime? **You could take a year to teach your dog some new tricks, or you could take a year to get rid of the negative portion of your mind, your orb. This we call free will. Judgment is only the assignment of relative importances to things.** Technology exists on our planet for getting rid of amnesia, negative, self-sabotaging attitudes and machines like the orb. Carlos Castaneda developed a technique he called "recapitulation." The purpose was to release energy knots relating to past experiences or interactions. The method is to move backwards re-experiencing the past occurrences and releasing the energy that has remained stuck with them. He claims to have had some success with the technique but he wasn't able to teach others how to use it.

Are you going to do something to better your state of mind before you make the next transition to the other side, or are you just going to "enjoy life" as they say on into oblivion?

57) CAN AN ORB THINK, FEEL AND PERCEIVE?

No, it does not have a viewpoint like you do. It thinks like an animal thinks, robotically. It is a machine, a quantum computer. It can see and respond to life, but only on a reactive, automated level. It only has the attitudes of the personalities that were postulated into it and it doesn't even have a brain. People don't need brains to think and experience. The brain is more of a shock absorber and switchboard to the body's nervous system. This explains why genetic defects that are born without their cerebral cortex can think and have a memory.

There exists a rare, completely bafflingly medical phenomenon - which has until recently been concealed - called hydrocephaly. To the normal materialistic Western biologist, this condition is astonishing, to say the least. In hydrocephaly, a person's cranial cavity is filled almost totally with fluid, not with brain matter. There may be only 5% or so of the brain in there; typically just the small portion on the tip of the spine. The other 95% of the brain case is filled with fluid. Yet the individual may be as normal as you or I. Except, of course, that x-rays of his head will astonish all the doctors. A few years ago, for example, such a hydrocephalic individual graduated from a university in Great Britain, with a degree in mathematics. British news actually made a video documentary on this subject, and particularly on that individual.

Now that I understand about orbs, I look at life in a totally different way. When I see a neighbor and her son fighting, invalidating each other, evaluating for each other, calling each other horrible names, threatening each other, I don't see two beings with evil intentions like I used to. Now I see two unconscious beings who are basically good, who know everything, who are matter, energy, space and time production units, who have awesome power and who are full of faith and truth being overwhelmed by their orbs. The beings have not been taking responsibility for each other and they have been building up mass in their orbs for trillions of years fighting each other. And they have no clue! They think they are doing the right thing and the best they can in a horrible situation because they don't realize that it is the orbs that are doing the fighting, not them! This is a war of the orbs, not of the spirits who are full of light and truth.

When I was younger I had no idea what to do about family squabbles. But now I know. I have to handle the orbs, not sympathize with my siblings. So I keep the combatants apart until the orb settles down. Then I talk to the siblings and try to get them to take responsibility for controlling the orb. This really works. Humans don't understand this and they suspect the intervention of the hand of God. And that is exactly correct; the only healing agency in the Physical Universe is the spirit.

At the moment of death, a being rises out of its host body. It takes its orb with it. It knows immediately that it has been set free and is going home (It is time to report back). These advanced beings need no one to greet them. However, most beings are met by guides and loved ones. This photo of the emergency entrance to the Kaiser Permanente Hospital in Hollywood, CA shows eleven beings thought to be attending a departing spirit.(below)Their purposes and activities have not been fully researched.

The disembodied being can "see" from its viewpoint in every direction at once. It can perceive a range of perceptions much greater than a human. It is not limited to the visible electromagnetic spectrum that humans can experience. It has an orb and a past that influences its behavior.

A cosmic being can enjoy existence and emotional impact; it can make plans and carry them out. It is very high esthetically and devotes most of its time to esthetics (like crop circles, pyramids, and other grand art forms). Spiritual beings associate socially with other beings and have a high sense of justice.

A cosmic being can be rendered unconscious by wave action; it can be hypnotized; it can be put to sleep; it can be made to forget its identity, just like humans. When a hypnotist puts a person in a hypnotic trance, it is the body which goes to sleep, not the spirit. The body just follows orders from the cosmic being or orb that is running it. But the spirit is so irresponsible that over 99% of the control comes from the orb and less than 1% is the spirit. Under hypnosis, it is the hypnotist that is dominating the cosmic being instead of the orb. The cosmic being can no longer control the body because it has gone into total agreement with the operator under hypnotic rapport. The operator's reality becomes the spirit's reality. When the hypnotist tells the body to raise its right hand, the cosmic being is confounded as he watches his right hand rise up. When the hypnotist tells the man that his daughter is not present on stage, the man cannot see her although she is standing directly in front of him. When the hypnotist asks the man to **read the serial number on the watch which is placed behind the daughter's head**, he does so even though he has to look through his daughters head to read it. When regressed to a past-life incident where the man was reading a newspaper, he can read it word for word even though it was a paper from the last century. When asked to read an ad from the **back side of that page that he never did see in a past life, he can read that ad word for word as well**. He could have total knowingness if it weren't for his orb. Cosmic beings have fantastic powers of perception. The kicker is; so do you! You are a cosmic being in a body.

Orbs can be made visible by certain electronic flows of wavelengths that are just now being invented on earth as solid state devices. In the distant past, wars between electronic beings and human-like creatures created unimaginable grief. Human-like creatures, trying to inhabit an area dominated by electronic beings, are commonly resisted and fought by the beings. The human-like creatures, fallen away from being free, and trapped within meat bodies, devise means to trap and harass the free beings. When the electronic beings have been reduced to little or nothing in power, they are used to motivate new bodies which are then physically controlled and taxed for the benefit of others.

Human-like creatures, suffering from amnesia, attack electronic beings with orbs who menace them (spook them, destroy their crops, scare their animals or harm their property). The electronic beings with orbs can kill humans or animals by throwing a strong electrical charge at them and they can scratch and cut the human's skin. See MUHNACHWA above. Thus wars develop between electronic beings and humanoid creatures. With the proper electronics, humanoid creatures can actually win the war.

58) CAN AN ORB BE DAMAGED?

A disembodied being associated with an orb can feel pain. But to damage an orb, it would require enough energy to nuke half a city. Orbs are very resilient. Beings can fly to the Moon to see if NASA really did land there, check out the rings of Saturn to see what spaceships are in standard orbit or check out the pyramids on Mars. Electronic Beings seem to live on our moon in large cities that have been observed by the Lunar Orbiter and the Clementine spacecraft. They usually try to stay on planets distant from humans. This data can be seen in my book, *Extraterrestrial Artifacts and the Secret of Human Identity.* The Surveyor 1 spacecraft made a soft landing on the Moon on June 2, 1966 near Flamsteed Crater in Oceanus Procellarum. It sent back a picture of a being's orb which turned out to see what fell into its back yard. NASA says the orb, which eclipsed the rim of a small crater, was simply a television ghost. JPL of Pasadena has created a NASA web site that details all the strange anomalies that have been seen on the Moon since the days of Galileo. There have been strings of moving lights, colored beacons, changes in the landscape and artifacts imaged by the Lunar Orbiter.

Orbs have also been associated with agriglyphs or crop circles. They have been reported to circle fields near Silbery Hill in England and cause the crops to lie down in huge aesthetic patterns. I have seen the video of two orbs circling a field at night and laying down an awesome pattern 300 feet in width. I have also heard reports and seen videos of large glowing orbs moving in the fields during day-light. The orbs vary greatly in size and power.

59) THE MECHANICS OF AGRIGLYPHS

There has been a lot of confusion about agriglyphs among humans. Some think they are transmitted signals from aliens, while others consider them to be 'graffiti' placed by 'passers-by' wanting to 'play on the sidewalk.' But most humans think they are practical jokes created by Doug and Don, the hoaxers, to get attention, and the farmers are furious with them. The Reptilians try to play down the crop circles and make everyone think they are hoaxes so they

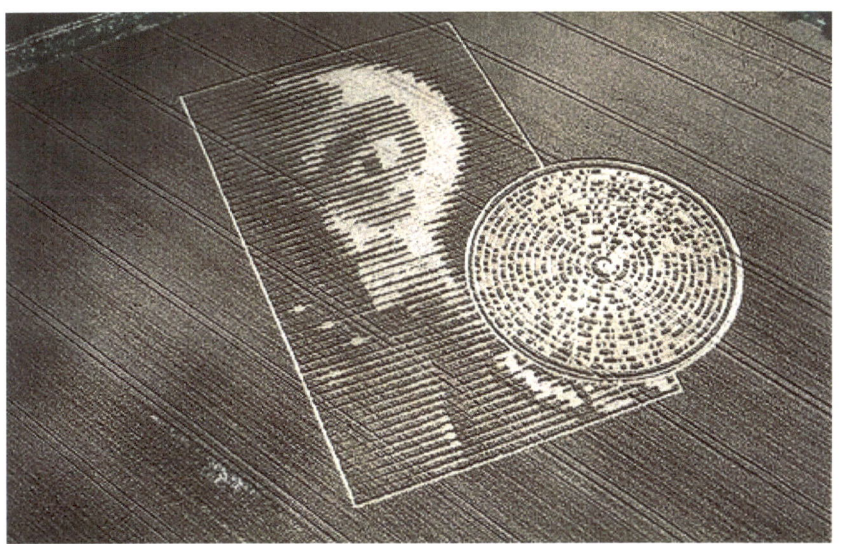

won't get any attention. When humans stay up late at night to watch the merkabahs come as balls of light and cause the rye grass to lie down in spectacular patterns, they are reminded of their spiritual heritage. This tends to wake up humans to the consternation of the Reptilians.

Dr. F. Lee Aeilts

This is the Crabwood agriglyph, found near Chilbolton, England, depicting an alien life form and a digital disk. The disk has been decoded by **Paul Vigay** as a message from the inside track spiraling out in ASCII format (a computer language used by the keyboard to communicate with the central processor). The message reads: **"Beware the bearers of FALSE gifts & their BROKEN PROMISES. Much PAIN but still time. (Damaged Word which could be BELIEVE). There is GOOD out there. We OPpose DECEPTION. Conduit CLOSING (BELL SOUND)"**

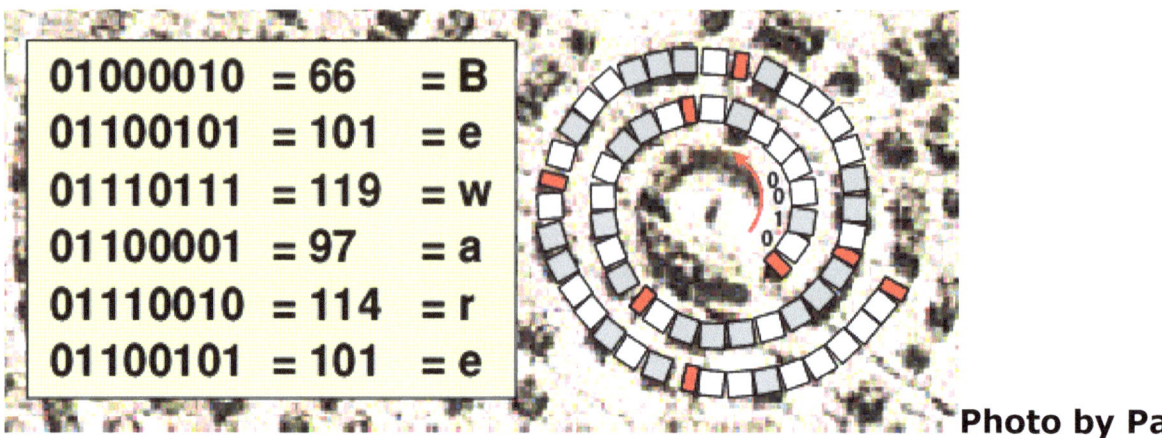

Photo by Paul Vigay

From my knowledge of the capabilities of orbs I can imagine a mechanism by which beings can mysteriously create vast crop-circles. Painting a picture on a cornfield is much like painting a picture on a computer using a paint program in the bit-map mode. This is the simple mode where each pixel or paint-able spot on the screen is either white or black. This would correspond to the cornfield where each corn stock would be up or down flat. It is also the same technique used by promoters at football games where the fans in the stands hold large colored sheets over their heads in a pattern so that the TV cameras can see the image of their mascot in the pattern.

So this is how it would work. Ten thousand beings would fly out to a pre-selected field and take positions at the base of each corn stalk. The commander would position himself at about 1,000

feet above ground level and give orders to the individual orbs to bend or not depending on their position in the desired pattern. You have to understand that a spirit can *be* anything that it can conceive whether it be a human, a dog, cockroach or blade of grass. As a life-static, the spirit can bend the corn stock as easily as the arm of a man or a metal spoon. (Much the same as Uri Geller would do it.) On command, each being does its job and lays down the crops in the desired manner. Then they all fly up to see how the pattern turned out. This explains the mysterious lights seen in the fields during the nights of the creation of agriglyphs. **This is the kicker; you can create these wonderful aesthetic effects too!** Just get rid of your orb!

60) HOW DO COSMIC BEINGS COMMUNICATE?

Cosmic beings communicate by telepathy. They work in the thought universe of images. They can throw an energy flow at things to move them. They can also put out tractor waves that pull things into them. They can travel at the superluminal speed of thought and they are not limited by atmospheres or gravity or temperature ranges like humans are. The beings quarrel with each other by showing each other disturbing pictures or by flowing energy at each other, but they are not very quarrelsome. Cosmic beings do not die and they have no gender. When cosmic beings decay they lose all their ability to use force so they take a human body and use the force of the physical body. After many lifetimes in animal bodies, building up their orbs, they pass out in the form of a solid, apathetic slumber. They have about as much life in them as a burned out cinder. They started out with the power of a super-nova. What a catastrophe! Don't let this happen to you! Get rid of your orb. The orb you see in photos following people around is the debilitating mental machine of the person it is following. It isn't some other being spying on someone. A friend of mine told me, "My wife has an orb and it follows her around everywhere she goes!" He didn't understand that the little orb was his wife's creation. It is near her because it is partly in her universe and partly physical. Her universe extends out to infinity but the orb is in the immediate vicinity because that is where mental masses form by the body's violent impacts with the Physical Universe.

61) THE MECHANICS OF ORB CREATION

Does a heavenly father and mother give birth to orbs? We see lots of orbs in our photographs so they must originate from someplace. Do they utilize a process similar to mitosis in cell division? No. ORBS are not sexual beings. Sexual propagation is the means that animals use to reproduce and start a new generation. To get a new spirit, all a life-static has to do is duplicate itself. Life-statics have this property. They can make a copy of themselves. If God were ever lonely he could duplicate himself as many times as he wanted and have a real party. Then a new life-static can mock up a ball and fill it with pictures and souvenirs of his travails in the Physical Universe and presto! You have a new orb. The life-static itself cannot be seen or photographed. It is a nothingness of absolute stillness and it contains no matter, energy, space or time, although it can generate these things. The pictures and structure of the orb does contain 21 grams of mass and this is what you see in the pictures of orbs. The density of an orb is greater than the density of the air that it travels in so a flash from a digital camera is partially reflected off the orb in the same way we discussed a wave traveling down a rope that is tied to a heavier rope is partially reflected.

The only thing that will get rid of an orb is to see it as it is through live communication. Now I finally know what it means to be responsible. It was my irresponsibility that built my orb! I cannot blame anyone else for creating it. I have to have the ability and willingness to assume that I caused everything in the universe. I cannot pass the buck. I can take responsibility. It is simply awesome to come to an understanding that we have such divine powers! Who would have guessed that we could do such a thing on this degraded planet? The truth has been denied us for so long!

Now that I am responsible I am not ridging up and creating my orb any more. I'm actually seeing how I have been self-sabotaging myself and now my orb is getting weaker; I am getting stronger. The road to higher states of awareness starts when a person has a dim little dawning of responsibility for his own existence. We cannot keep going downhill like we have in the past. I realize that I have had something to do with the bad luck I've had in life. My future depends on increasing my ability to confront and cope with existence; not on raising my income and improving my lifestyle. Those are temporary conditions and will pass away. But my abilities and knowledge will create a wonderful future for me and my friends. Also, all security derives from knowledge.

Man's only real sin is the betrayal of self. A man betrays himself; he is faithless to himself and that's how he builds his orb and destroys himself. If you're not true to your basic purpose you'll die spiritually and build an orb. Every pain, unconsciousness, blackness, ridge, and picture in an orb was caused by a person altering his basic purpose.

62) HOW COSMIC BEINGS GET BODIES

A human body is a vegetable. It is not even a sentient vegetable because it lacks perception in the whole range of human experience. Like any vegetable it grows from seed and its habit patterns help it survive. All vegetables get used by others one way or another and a human body is no exception. A meat body, whether of an ant or a human, is still an animated vegetable. By itself the body would live, run around, react, sleep, kill prey and live a life no better than a field mouse or zombie. Put a cosmic being in charge of it and it becomes ethical, moral, esthetic, goal oriented, intellectual, wears fashionable clothes, shaves, flushes the toilet and goes to church. It becomes this strange entity called Homo sapiens.

Cosmic beings are attached to bodies by connecting to the left and right motor control centers of the cerebral cortex. There is also a biophotonic connection to every cell in the body. Once attached, no other being can get in to take over the body. Sometimes a group of beings decide by agreement to use one body. In this case, one being at a time is at the controls of the body and the other cosmic beings are not conscious of the body's perceptions although they can watch the actions of the body from their exterior positions (viewpoints). An example of this is Shari the Extraterrestrial. She works with two other cosmic beings who are commanders on a space ship in a near Earth orbit. Channelers in general are other examples of this phenomenon. This gets very New Age.

A life-static, or cosmic being, doesn't need an orb or a meat body or motors of any kind. It is a matter, energy, space and time production unit. It is a perpetual motion machine and would live forever if it weren't corrupted by its association with the Physical Universe. It is impossible to harm a cosmic being. Only its own considerations could inhibit it. Only by attaching the life-static to masses (such as orbs, bodies, vehicles, etc.) and then damaging those masses can we get a being to think it has been harmed. But the truth is, a being of no dimension, no wavelength, no matter, energy, space or time cannot be harmed no matter how much excruciating pain it pretends to exhibit. It cannot be a victim, and it will never recover its cosmic beingness until it realizes that it is the cause of any difficulties it has ever had. The only effects it can experience are those which it dreamed up to experience. No one else caused it to participate in this Universe. Just think about it. Who made all the decisions to land you in the place where you are right now? You are cause! You cannot say that you didn't choose your parents. You cannot say that you didn't choose your mate or the city where you live. You cannot point your finger at anyone else.

63) HOW TO PHOTOGRAPH ORBS

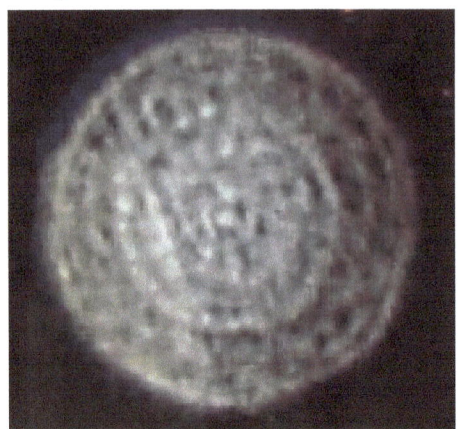

It is easy to photograph orbs if you own a digital camera with a built-in flash. But be careful of false images like these three. On the left is the image of a raindrop. Orbs do not exhibit patterns of concentric circles. You cannot see forms within them because they are composed of very fine quantum waves and are translucent to visible light. A solid orb, or dust orb, is created because a reflective solid airborne particle, such as a dust particle, is situated near the camera lens and outside the depth of field, in other words out of focus. The pinpoint of light reflected from the dust particle that would be seen if it were near the focal distance grows into a circle of confusion with increasing distance from it. Also, be careful of solar reflections like the one at the right. The photographer noticed a little disk of light on the camera lens which moved when the sun's angle changed. It was a solar reflection in the camera's lens system.

To get the best pictures, the background light level should be low. That causes the camera to open up to its maximum light sensitivity. The flash sends out a light pulse of roughly 1/1,000 of a second that reflects off the surface of the orb, revealing its presence and showing its details. If the background is dark, as with a night sky, you can get good contrast to maximize the details. In the daylight shots, the orbs are very translucent and you'll probably only see a thin bubble in the photo. The biggest secret to tracking orbs is to take lots of pictures and erase all the ones which don't show anything. The photo below is not an orb. It is a picture of a flash going off. Eventually you'll discover where the beings hang out or which human is trailing external orbs so that you'll get one or more orbs in every shot. It took me a couple of days of sleuthing at night in my neighborhood before I realized that disembodied beings were visiting the Kaiser-Permanente hospital across the street regularly for the purpose of welcoming dying humans into the thought universe or finding new bodies for beings in the maternity ward.

64) WHY IS IT BENEFICIAL TO UNDERSTAND ORBS?

You can use the knowledge of orbs to make otherwise impossible strides in your understandings of living beings. You are living with paradoxes which no philosopher in all the ages ever reconciled. The problem of evil in the world, the injustice of death, man's inhumanity to man, the liability of assisting another, the problem of loving others and yet surviving yourself, the penalty of being kind and merciful, and all the other religious paradoxes are solvable with this knowledge. Without knowing the difference between the spiritual and the physical attributes of a human being you are likely to demand of a spiritual/material composite being that he/she be self-determined when every slightest motion from a hard universe can wipe him/her out and send him/her into unconsciousness. You are likely to demand that he/she be "careful" when his/her only salvation is to be carefree. Without this data you are saddling yourself with all the unanswered riddles of an aberrated human condition. And you are condemning yourself to eventual spiritual extinction as a being as you descend the dwindling spiral of decay in bodies because the life-static, as part of the composite being, decays fast and soon dies forever in the rigid apathy of matter, energy, space and time. Without this data you have no hope of rehabilitation. **It applies to all humans on this planet**. We have no shortage of beings to save, and we have only a narrow window of opportunity now as our planet catches its breath between wars and calamities.

65) BIOPHOTONICS – QUANTUM CONTROL OF BODIES

A **biophoton** (from the Greek βιο meaning "life" and φωτο meaning "light") is a photon of light emitted in some fashion from a biological system. From a scientific point of view, there is no difference between such a photon and a photon emitted by any other physical process. All live plant and animal cells continuously exchange biophotons. The first systematic research into the role of light in living processes was done by the Russian scientist Alexander Gurwitsch (1874-1954) in the 1920s. Gurwitsch established as a conclusive hypothesis that every living cell, plant or animal, emits light at a very weak level. This is what he found:

- *The roots of two onions are positioned perpendicularly so that the tip of one root points to one side of the other root. Gurwitsch found that there was a significant increase in cell divisions (mitosis) on this side, compared to the opposite, "unirradiated" side. The effect disappeared when a thin piece of window glass was placed between the two roots, and reappeared when the ordinary glass (which is opaque for ultraviolet light) was replaced with quartz glass, which is transparent for ultraviolet light.* Gurwitsch termed this effect Mitogenetic Radiation.

- Gurwitch's approach to biology came under fierce attack by the proponents of genetics and molecular biology, which then, after the war, was made the dominant field of research. The major point of criticism from these circles was, that the mitogenetic radiation did not exist at all, or, if it did exist, it had no biological relevance whatsoever.

SCHEMATIC OF A. B. BURLAKOV'S EXPERIMENTS WITH FISH EGGS AND MITOGENETIC RADIATION IN RUSSIA

Samples of fertilized fish eggs in different phases of development were brought into optical contact with each other. Burlakov found that if the age difference between the eggs or larvae was not too large, there was a significant acceleration in the development of the younger eggs relative to the older ones. However, if the age difference was large, the younger eggs showed a strong retardation in development; even deformities and higher death rates occurred. When Burlakov used normal window glass as a filter, all these effects disappeared, but with quartz filters, the effects could be observed.

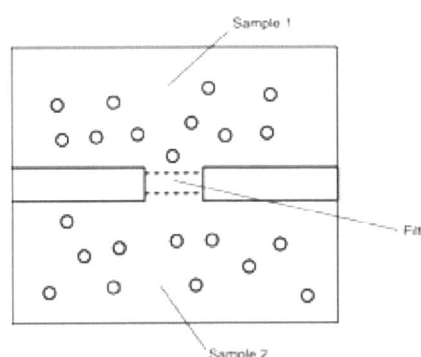

In Gurwitsch's time, it was technically impossible to directly measure the weak light emission from cells. This became possible only in the 1950s, when a group of Italian astronomers developed a very sensitive photo-multiplier, (right) which they used to make distant stars visible. When used on biological samples, it was shown that leaves, germs of wheat, corn, beans, and so on, emit a constant, but weak light. These results created a brief uproar in the West, but the affair was then essentially forgotten.

Intercellular photons are the communication system that every one of the cells of your body uses to communicate to all the other cells and to the world around you. Achieving a certainty on this knowledge will give you a fantastic advantage over other beings who don't understand it and who live in the darkness of twenty-first century psychology. This data is not NEW AGE or religion. It is hard science developed by painstaking research and published in peer-reviewed journals.

66) DR. FRITZ-ALBERT POPP

Dr. Fritz-Albert Popp's popular books in German, *The Biology of Light* and *Biophotons, the Light of Our Cells*

COSMIC MECHANICS

With Popp's photomultiplier machine, it was possible to prove beyond any doubt that low-level light emissions are a common property of all living cells. It has different intensities for plant or animal cells, for different cell types, and it can vary from one moment to the next. It is not regular, but comes often as "photon explosion" (spikes), especially when the cells are irritated by outside means. Dr. Popp's group also discovered that the photonic radiation is not the product but essentially the initiator of chemical reactions in the cells. The radiation submits the information within and between cells.

Popp demonstrated the biophoton nature of the global aura with two quartz test tubes containing living pig blood. An agent was added to one of the test tubes and as expected the pig's blood reacted by producing antibodies. Although no agent was added to the second tube, it nevertheless produced identical antibodies. This experiment was now repeated with a light-proof barrier inserted between the two quartz test tubes and the potential life-saving antibody information could not be communicated to the second tube. This elegant experiment clearly showed that biophotons can communicate the information needed to initiate the vital biochemical processes of life, through long-range-coherent electromagnetic coupling.

Popp thought he had discovered a cure for cancer. In 1970 he was examining benzo[a]pyrene, a polycyclic hydrocarbon, one of the most lethal carcinogens and illuminated it with ultraviolet light. He discovered that benzo[a]pyrene had a strange optical property. It absorbed light but then re-emitted it at a completely different frequency. This chemical acted like a biological frequency scrambler. Popp then performed the same test on benzo[e]pyrene which is virtually identical to benzo[a]pyrene with one tiny difference, an alteration in its molecular makeup. This tiny difference in one of the compound rings rendered benzo[e]pyrene harmless to humans. This particular chemical passed light right through the substance unaltered. Popp tested thirty-seven other chemicals and learned to predict which compounds were carcinogenic. In every instance, the compounds that were carcinogenic took the UV light, absorbed it, and changed the frequency. Curiously, each of the carcinogens reacted only to light at the specific frequency of 380 nanometers. Popp wondered why a cancer-causing substance would be a light scrambler. He discovered an undisputed scientific phenomenon called "photo-repair." Biological lab experiments prove that if you blast a cell with UV light so that 99% of the cell, including its DNA, is destroyed, you can almost entirely repair the damage in a single day by just illuminating the cell with the same wavelength of a very weak intensity. Popp also knew that patients with a skin condition called xeroderma pigmentosum eventually die of skin cancer because their photo-repair system doesn't work and so doesn't repair solar damage. Popp was shocked to learn that photo-repair works most efficiently at 380 nanometers - the very same wavelength the cancer-causing compounds would react to and scramble! This means that there must be some light in the body responsible for photo-repair. Cancerous compounds cause cancer because it permanently blocks this light and scrambles it, so photo-repair can't work any more.

Multiple Sclerosis was found to exhibit just the opposite effect. Individuals with this disease were taking in too much light. This inhibited the cells from doing their job. It is like too many soldiers marching in step across a bridge, causing it to collapse. MS patients were drowning in light.

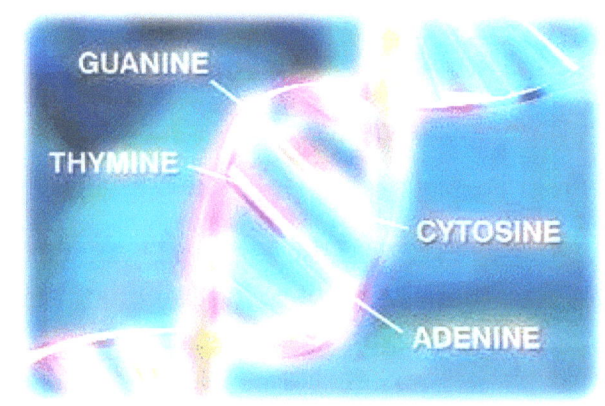

Popp also examined the effect of stress. In a stressed state, the rate of biophoton emission went up – a defense mechanism designed to return the patient to equilibrium. Popp had a model which provided a better explanation than the current neo-Darwinist theory for how all living things evolve on the planet. Rather than a system of fortunate but ultimately random error, if DNA uses frequencies of all variety as an information tool, this would suggest instead a feedback system of perfect communication through waves which encode and transfer information.

Popp came to realize that light in the body might hold the key to health and illness. He compared the photons in eggs from free range chickens to those raised in hen houses. The photons in the eggs produced by the free range chickens were far more coherent than those in the hen house chicken's eggs. He then used biophoton emissions as a tool for measuring the quality of food. The healthiest food had the lowest and most coherent intensity of light. Any disturbance in the system would increase the production of photons. Health was a state of perfect subatomic communication, and ill health was a state where communication breaks down. We are ill when our waves are out of sync.

Popp experimented with a number of non-toxic substances purported to be successful in treating cancer. All of the substances except one increased the photons from tumor cells, making them even more deadly to the body. Mistletoe seemed to help the body "resocialize" the photon emission of tumor cells back to normal. Popp tested mistletoe and various other plant extracts on samples of a middle aged woman's cancerous tissue and found that one particular mistletoe remedy created coherence in the tissue similar to that of the body. With the agreement of her doctor, the woman began foregoing any treatment other than this mistletoe extract. After a year all her laboratory tests were virtually back to normal. A woman who was given up as a terminal cancer case had her proper light restored just by taking an herb.

Popp didn't know about the orb. The orb is the greatest inhibitor of biophotonic communication and coherence in the human body. When a person receives a painful bruise such as a dog bite to the left arm, the orb records the biophotons coming from the source of the pain and the location where it occurred. Later when the person is in the same location or sees a similar dog, a bizarre pain in the left arm will be felt. The person will have no idea where the pain is coming from because the unsuspected orb is unknowingly playing back the biophotonic recording of the dog bite. This will cause skin resistance and biophotonic emission to increase. If, during counseling, the person realizes that the orb is playing back a recording of the pain, all the pain, high skin resistance and high biophotonic emission will immediately cease. The orb gets restimulated often during the day and breaks down inter- cellular communication and causes aches and pains. The best thing you can do to assure vibrant health and fitness is to get rid of your orb. Taking vitamins and minerals won't do it. The power of positive thinking won't do it, although Norman Vincent Peale will try to convince you that it will. Will power won't do it, notwithstanding Anthony Robin's books and tapes. Humans realize that those things don't work to maintain a consistently healthy lifestyle. We've all tried them and they have failed us.

Some of us have experienced miraculous healings. The healings occurred as a result of a modification of the person's orb even though the credit was given to a pastor, a guru, a saint, the Virgin Mary or Jesus. I don't mean to step on anyone's religious toe's here. If this isn't true for you, it isn't true. I don't want to invalidate your faith. I just want you to understand the mechanics so that you won't need faith to understand healing. You'll have a certainty. Then you can build from there by exercising faith in greater things, such as yourself as a cosmic being.

Extremely powerful methods have been developed to eradicate orbs. Carlos Castaneda's technique of recapitulation was one of the first techniques developed to annihilate standing waves in the cosmic being's universe. Since these standing waves are quantum entangled with the space resonances of the being's orb in the material universe, the corresponding space resonances in the orb are also annihilated. The purpose of Castaneda's technique is to release energy knots relating to past experiences or interactions. The method is to move backwards, re-experiencing the past occurrences and releasing the energy that has remained stuck with them. Recapitualation frees the energy entangled in painful memories and increases the person's clarity of thought.

More than 50 years ago Erwin Schrodinger made the startling proposal that life is based on quantum-mechanical principles. He pointed out that classical laws such as Newton's laws of gravity and motion are all statistical, true for collections of billions of atoms or molecules but invalid at the level of individual particles. He maintained that life was based on the dynamics of individual particles and therefore subject to quantum laws.

Johnjoe McFadden, Professor of Microbiology, University of Surrey, says that the living cell is nature's nanotechnolgy. "Just as engineers and physicists working at the nano-scale level must include and exploit quantum mechanics in their models, so evolution over three billion years must have incorporated quantum dynamics. Quantum mechanics is likely to be as fundamental to life as water. Indeed, recent experiments and simulations indicate that protons involved in hydrogen bonding in water are highly delocalised (that is, in a superposition of being in two separated locations). Hydrogen bonding is probably the most fundamental biochemical interaction, involved in DNA base-pairing, enzyme catalysis, protein folding, respiration, and photosynthesis. If quantum delocalization lies at the heart of this phenomenon then it is central to life. Indeed, several researchers (including myself and Paul Davies) have proposed that quantum mechanics may account for that ultimate of biological mysteries: the origin of life itself."

Now that we've established a sufficient purpose for studying biophotonics, let's get on with it. The science of biophotonics has some technical terms which can be confusing to the uninitiated, so full definitions will be given in this book where the terms are first mentioned.

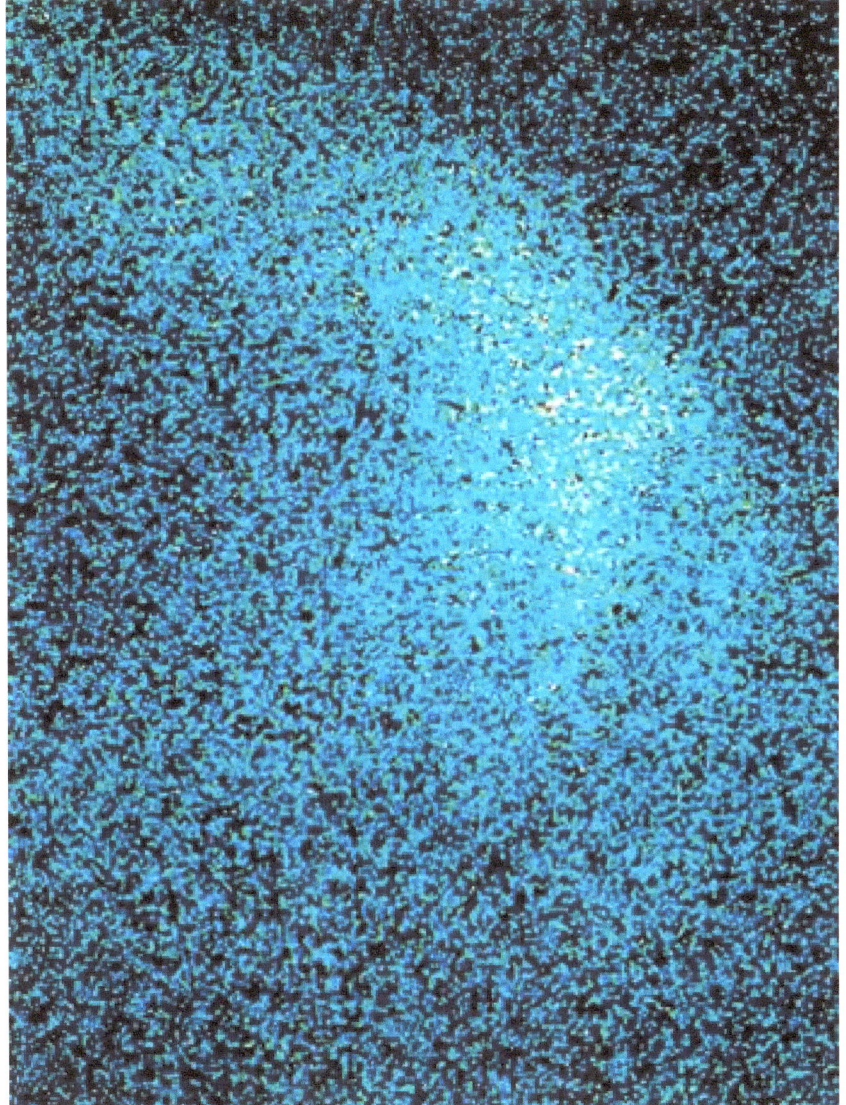

A **biophoton** (from the Greek βιο meaning "life" and φωτο meaning "light") is a photon of light emitted in some fashion from a biological system. From a scientific point of view, there is no difference between such a photon and a photon emitted by any other physical process. The term "bioluminescence" is generally reserved for higher intensity chemical systems such as fireflies and glowing fish while "biophoton emission" refers to the more general phenomena of low-intensity photon emission from living systems. All live plant and animal cells continuously exchange biophotons and the (accupuncture) meridians of the body are the major channels of communication. Above, an image of the first biophotons seen in the cells of a germinating soybean seed.

BIOLUMINESCENCE

The Pyralis firefly (also known as the lightning bug) is a common firefly in North America. This partly nocturnal, luminescent beetle is the most common firefly in the USA. It is an example of bioluminescence. Biophotonics is different. The term *biophoton* is used specifically to denote those photons that are detected by biological probes as part of the general weak electromagnetic radiation of living biological cells. **All plant and animal cells receive and transmit photons,**

even in total darkness, and these photons are the cell's broadband communication link to the rest of the body and to the entity or spirit controlling it. Biologists used to think that the brain was the mind and intracellular communication was done by chemical nerve impulses traveling between 10 and 100 meters per second. That type of communication does exist but it is not fast and doesn't carry much information. For an athelete to give an impressive gymnastic performance, he must communicate with his body at the speed of light with broadband communication. In order for organisms to survive in the material universe, they need fast intercellular communication and lots of it! The sources of the photons are the DNA and corresponding resonators in the cells.

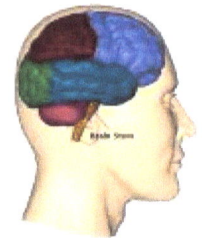

The brain is not the mind. The brain is the gray multi-folded mass in the head that operates like a telephone switchboard to the body. Doctors have noted that people born without the gray matter in their heads still have the ability to think and reason. The mind is the incorporal spiritual being (you) who controls the body with biophotonic quantum waves. The reactive mind is the little ball of quantum mechanical circuits which exist in the vicinity of the body and which can be photographed. It weighs approximately 21 grams and many researchers refer to it as an "orb." . It works on a subconscious, quantum mechanical level and is the source of all the psychosomatic illnesses of mankind. These circuits are the "invisible" storehouse of recorded painful experiences which react on humans like post-hypnotic commands. Getting rid of these circuits brings greater health, increased abilities, higher IQ, faster reaction time, enhanced creative imagination, amazing vitality, deep relaxation, improved memory, stronger will power, a magnetic personality and good self-control.

Everything in biophotonics is based on a very important principle. A person is a lot more than just his body or his mind. A person is an unlimited spiritual being. I'm not talking about "soul" here, which religions use to denote the entity that we have to save or that we could sell to the devil. No, I'm talking about you, the cosmic being who lives once in a succession of bodies. I'm going to bring back to you your own awareness of how you control bodies and I'm going to show you the quantum mechanical experiments that lead to these discoveries. This book is about YOU. It explains a workable way for you to recover your natural spiritual power and ability. It explains how you can use this knowledge to change the conditions in your life for the better—fantastically better!

This book doesn't soothe you with words like Wayne Dwyer or Carolyne Myss. I know that you feel wonderful after listening to their CDs, but your orb is going to kick in when you go back to work, experience road rage, handle your kids, talk to your psychiatrist, try to leave your lover, listen to the news or try to make a sale. This book is about getting to the root of the problem, the orb. When you can confront the orb; then kids, wives, lovers, employers, crazy drivers and life's slings and arrows won't get you down.

Going to church and listening to evangelical ministers won't handle the problem either. Billy Graham is wonderful, but after turning over your life and fortune to "God" you still have to confront crime, insanity, war, old age and disease. It is so much easier if you don't have an orb to shackle and torment you.

67) THE MECHANICS OF PHOTONS

The **photon** is one of the elementary wave structures. Its interactions with electrons and atomic nuclei account for a great many of the features of matter, such as the existence and stability of atoms, molecules, and solids. These interactions are studied in quantum electrodynamics (QED), which is the oldest part of the Standard Model of particle physics.

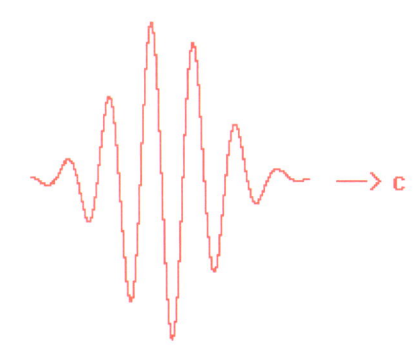

In many circumstances, a photon acts as a classical particle, for instance when registered by the light-sensitive device in a camera. In other circumstances, a photon acts like a classical wave, as when passing through the optics in a camera. According to the so-called wave-particle duality of quantum physics, it is natural for the photon to display either aspect of its nature, according to the circumstances. Normally, light is formed from a large number of photons, with the intensity related to the number of them. At low intensity, it requires very sensitive instruments, such as photo-multiplier tubes to detect the individual photons. Some say a photon "sometimes acts like a wave and sometimes acts like a particle". This is slightly misleading, because a photon *always* acts like *both*. For example, when shooting single photons through a slit, a detector can detect each photon when it hits—but over time, the detector will detect the same diffraction pattern as it would if the photons were given off all in one burst.

The photon can be seen to behave as a wave or a particle, depending on how it is measured. But remember that the particle properties of photons are the result of the wave structure of matter. So basically, all quantum particles are waves because there is no such thing as a discrete solid chunk of anything in the Physical Universe. It is all waves.

PROPERTIES OF PHOTONS

Photons are commonly associated with visible light, but this is actually only a very limited part of the electromagnetic spectrum. All electromagnetic radiation is quantized as photons: that is, the smallest amount of electromagnetic radiation that can exist is one photon, whatever its wavelength, frequency, energy, or momentum, and that light or fields interact with matter in discrete units of one or several photons. We learn from quantum mechanics that photons are fundamental wave structures.. They can be created and destroyed when interacting with other wave structures, but do not decay. This means that if you set up a quantum mechanical circuit in your reactive mind with photons, that circuit will continue to affect you for trillions of years after conditions have changed and you no longer have a need for it. The circuit will float in time forever and bug you until you destroy it by interacting with it with new quantum energy. For example, if a hundred years ago, in another lifetime, you died in the Titanic disaster and decided then that you'd never go near boats again, that decision, in quantum mechanical form, will affect you in this life. Going aboard a boat will give you a bizzare anxiety that you and your psychiatrists won't understand. Remembering the incident and examining it carefully will erase it and cause the bizzare reactions to disappear. This happened to a friend of mine who I was counseling. He was afraid of boats. When his family took a trip from England to the Portugues Azores, he got into a fight with another boy on the deck of the ship. The boy's mother broke up the fight and threw my friend up against the fence at the edge of the deck. This action caused him to fear being thrown overboard and drowning. He was shaking as he made his way down into the lower decks to avoid falling overboard and drowning. In a session that I had with him, he discovered that his overwhelming fear came from a previous life where his ship had broken on some coastal rocks in the North Atlantic and he had jumped overboard to save his life. The surf against the rocky shore was impossible to fight so he tried to get back to the ship amidst the flotsam and jetsom. His body soon tired and went numb in the cold North Atlantic water so he died in the attempt. Upon realizing that this event was the source of his fears about boats, he immediately was relieved and the fear was gone. This phenomenon was discovered by Dr. L. Ron Hubbard, author of *Dianetics, Modern Science of Mental Health*. (ISBN 0-88404-632-X)

Planck's Constant and the Energy of a Photon

In 1900, Max Planck was working on the problem of how the radiation an object emits is related to its temperature. He came up with a formula that agreed very closely with experimental data, but the formula only made sense if he assumed that the energy of a vibrating molecule was **quantized**--that is, it could only take on certain values. The energy would have to be proportional to the frequency of vibration, and it seemed to come in little "chunks" of the frequency multiplied by a certain constant. This constant came to be known as **Planck's constant**, or *h*, and it has the value

$$h = 6.626 \times 10^{-34} J \cdot s$$

It was a very small value and it was an extremely radical idea to suggest that energy could only come in discrete lumps, even if the lumps were very small. Planck actually didn't realize how revolutionary his work was at the time; he thought he was just fudging the math to come up with the "right answer," and was convinced that someone else would come up with a better explanation for his formula. Einstein took his data at face value. Based on Planck's work, Einstein proposed that light also delivers its energy in chunks; light would then consist of little particles, or **quanta**, called **photons**, each with an energy of Planck's constant times its frequency. In that case, the frequency of the light *would* make a difference in the photoelectric effect. Higher-frequency photons have more energy, so they should make the electrons come flying out faster; thus, switching to light with the same intensity but a higher frequency should increase the maximum kinetic energy of the emitted electrons. If you leave the frequency the same but crank up the intensity, *more* electrons should come out (because there are more photons to hit them), but they won't come out any *faster*, because each individual photon still has the same energy. And if the frequency is low enough, then none of the photons will have enough energy to knock an electron out of an atom. So if you use really low-frequency light, you shouldn't get *any* electrons, no matter how high the intensity is. But if you use a high frequency, you should still knock out some electrons even if the intensity is very low. Therefore, with a few simple measurements, the photoelectric effect would seem to be able to tell us whether light is in fact made up of particles or waves.

Dr. F. Lee Aeilts

In 1913-1914, R.A. Milliken did a series of extremely careful experiments involving the photoelectric effect. He found that all of his results agreed exactly with Einstein's predictions about photons, not with the wave theory. Einstein actually won the Nobel Prize for his work on the photoelectric effect, not for his more famous theory of relativity.

Some experimental results, like this one, seem to prove beyond all possible doubt that light consists of particles; others insist, just as irrefutably, that it's waves. Now we know that all atomic 'particles' are actually space resonances or standing waves that exhibit the characteristics of discrete particles.

A photon of a definite frequency is not a localized wave structure. You cannot pin it down to existing in any particular place. It exists everywhere but has a probability of showing up in a specific point in time and space. It is the link between quantum waves and the Physical Universe, so it is part static, (nothingness) and part dynamic (somethingness). The interplay between the static and the dynamic results in the world we see. Just as a spiritual being has no mass, energy, space, time, size, form or location, photons and quantum waves exhibit a position-frequency uncertainty relation. According to the quantum electrodynamics of the Standard Model, photons have zero rest mass and zero electric charge, but they do carry energy, momentum and angular momentum. Although the photon is generally accepted to be massless, experiments may only show that its mass is consistent with zero. As massless particles, photons must always move at c (often called the speed of light in vacuum), which is approximately 300 million meters per second..

Because of special relativity, photons always move at a constant speed with respect to all observers, regardless of the observers' own velocities. The energy and momentum carried by a photon is proportional to its frequency (or inversely proportional to its wavelength) with a constant of proportionality equal to the Planck constant. The momentum carried by a photon can be transferred when it interacts with matter. This momentum can be used for a space drive engine.

(following) The force due to photons interacting with a surface is called radiation pressure, which may be used for propulsion in space as with a solar sail.

COSMIC MECHANICS

Photons are deflected by a gravitational field twice as much as Newtonian mechanics predicts for a test mass traveling at the speed of light. This observation was a key piece of early evidence supporting Einstein's theory of gravitation, general relativity. In general relativity, photons (as well as any other object in a free fall) always travel in a "straight" line, taking into account the curvature of spacetime.

CREATION OF PHOTONS: Photons are created in profusion by spiritual beings like you and me. Each cell of our bodies receive and transmit photons, biophotons. You can image them with digital cameras. (below)

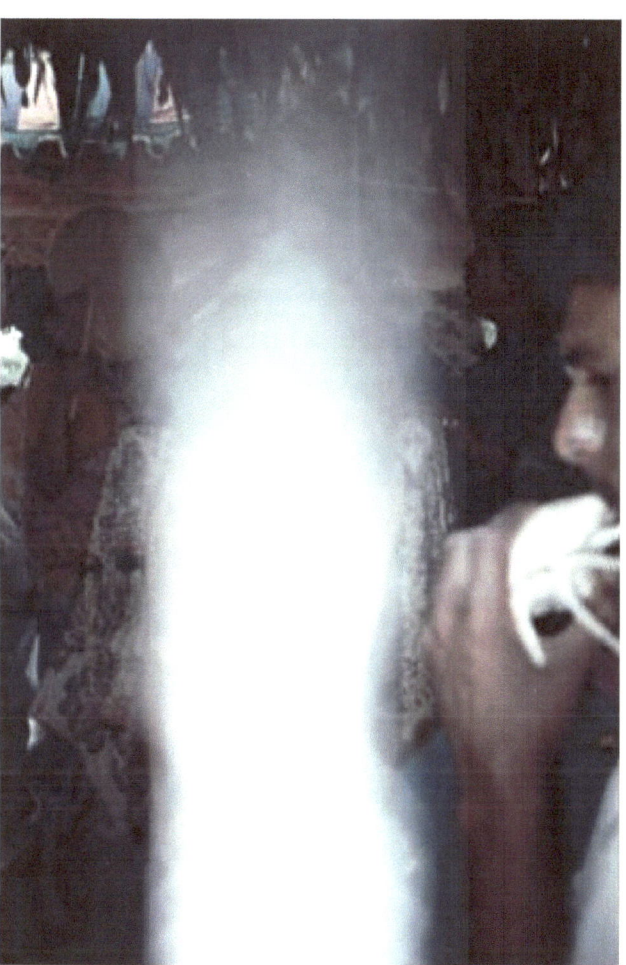

Photo taken the day before Shri Swamiji's body was interred shows a light body obscuring the body.

Photons are also produced by changes in quantum state of a charged particle, from a state of higher energy to a state of lower energy, for example by atoms when a bound electron moves from one orbital to another orbital that has less energy or when a free electron becomes bound by an atom. **Every time a chemical bond forms, photons gets trapped, and every time a**

chemical bond breaks, photons gets emitted. That's been known for a long time. Examples are fire, light bulbs, fluorescence, sunlight, lightning and afterburners.

Photons can also be emitted by an unstable nucleus when it undergoes some types of nuclear decay, producing gamma radiation. The frequency of an emitted photon is the result of a *beat* between the frequencies (energies) of the two stable states involved in the quantum transition, and the photon may be viewed as the electromagnetism of the moving charge involved in the mixture of states during the transition; the photon frequency or energy therefore represents the difference in energies of the states. If a mixture of two states does not result in an oscillation of the electric charge (mean probabilistic position), then the transition is said to be "disallowed" as there will be no coupling to radiation. Photons are also radiated whenever charged particles are accelerated, as in the production of bremsstrahlung (x-ray tubes), cyclotron or synchrotron radiation.

68) THE MECHANICS OF ATOMIC SPECTRA

Certain gases at low pressure emit light in a set of *discrete bands* of the electromagnetic spectrum. This observation had to be accounted for. This is quite different from the radiation emitted by solids, which is spread evenly across the electromagnetic spectrum. The radiation emissions of these gases were important because they showed that at least under some circumstances, the orbits of the electrons could not be at just any distance from the nucleus, but were confined to discrete distances (or *energy states*).

A. Continuous spectrum and B. line spectrum of hydrogen [3]

If the electrons in these gases were free to orbit at any distance, then the light emitted from them would have been spread evenly across the electromagnetic spectrum. Instead, what experimenters saw was that the light from these gases showed a distinct *line pattern*. That is to say that the light being emitted was only seen in a certain set of wavelengths, with empty spaces in between.

These line-spectra were different for each gas, and was found to be the characteristic of its atom. Today, astronomers use line-spectra to detect the elements present in stars.

Radio, television, radar and other types of transmitters used for telecommunication and remote sensing routinely create a wide variety of low-energy photons by the oscillation of electric fields in conductors. Particle-like properties are not detectable at such low frequencies. Magnetrons emit coherent photons used in household microwave ovens. Klystron tubes are used when microwave emissions must be more finely controlled. Masers and lasers create monochromatic photons by the stimulated emission process. More energetic photons can be created by nuclear transitions, particle-antiparticle annihilation, and in high-energy particle collisions.

ANNIHILATION OF PHOTONS

Since there is no conservation of photon number, photons may be annihilated. This process may occur as the time reversal of any way in which photons are created; for example, in absorption by atoms, nuclei or molecules. As long as total energy and momentum are conserved (say, by transfer to an optical medium), a photon may split into two photons, or two photons can combine into one.

SPIN

Photons have spin 1. Other particles have spin 0 or -1. There exist only three spin states, 0, 1 or -1.

QUANTUM STATE

Visible light from ordinary sources (such as the Sun or a lamp) is a mixture of many photons of different wavelengths, phases and polarizations. One sees this in the frequency spectrum, for instance by passing the light through a prism. In so-called "mixed states", which these sources tend to produce, light can consist of photons in thermal equilibrium (so-called black-body radiation). Here they in many ways resemble a gas of particles. For example, they exert pressure, known as radiation pressure.

MOLECULAR ABSORPTION

A typical molecule, M, has many different energy levels. When a molecule absorbs a photon, its energy is increased by an amount equal to the energy of the photon. The molecule then enters an excited state, M^*.

$$M + \gamma \rightarrow M^*$$

PHOTON MASS AND PHOTONS IN A BOX Although a single photon has zero mass, multi-particle objects including photons may collectively have mass. For example, a mirrored box containing a gas of photons, or even a single photon, with total energy *E* will have greater mass (by $\Delta m = E/c^2$) than if the box did not contain photons.

PHOTONS IN VACUO

In empty space (vacuum) all photons move at the speed of light, *c*. The meter is in fact *defined* as the distance travelled by light in a vacuum in 1/299,792,458 of a second.

According to one principle of Einstein's special relativity, all observations of the speed of light *in vacuo* are the same in all directions to any observer in an inertial frame of reference. This principle is generally accepted in physics, since many practical consequences for high-energy particles in theoretical and experimental physics have been observed.

Since photons move at the speed of light, by relativist time dilation they do not take any time to get from their source to where they are finally absorbed; that is, they have zero lifetime but can travel arbitrarily far. The emission and absorption events are at zero space-time interval. From this point of view, first articulated by Gilbert N. Lewis in 1926, **the photon's energy never exists in the vacuum, but transfers from the source to the absorber without delay. This is a strange, almost spiritual, fact. It is due to the fact that the photon is part static (spirit) and part dynamic (physical).** So when starlight leaves its galaxy 800 million light years away it doesn't exist again until it is detected by the retina in your eye.

PHOTONS IN MATTER

When photons pass through matter, such as a prism, different frequencies will be transmitted at different speeds. This is called dispersion of colors, where photons of different frequencies exit at different angles. A similar phenomenon occurs in reflection where surfaces can reflect photons of various frequencies at different angles.

The associated dispersion relation for photons is a relation between frequency, f, and wavelength, λ, or equivalently, between their energy, E, and momentum, p. It is simple *in vacuo*, since the speed of the wave, v, is given by

$$v = \lambda f = c$$

The photon quantum relations are:

$E = hf$ and $p = h/\lambda$

Here h is Planck's constant. So one can also write the dispersion relation as

$$E = pc$$

which is characteristic of a zero-mass particle. One sees that Planck's constant relates the wave and particle aspects.

In a material, photons couple to the excitations of the medium and behave differently. These excitations can often be described as quasi-particles (such as phonons and excitons); that is, as quantized wave- or particle-like entities propagating though the matter. "Coupling" means here that photons can transform into these excitations (that is, the photon gets absorbed and medium excited, involving the creation of a quasi-particle) and vice versa (the quasi-particle transforms back into a photon, or the medium relaxes by re-emitting the energy as a photon). However, as these transformations are only possibilities, they are not bound to happen.

A biophoton is very weak. This amount of light has been compared to that observed from a candle viewed at a distance of 10 kilometers. The detection of these photons has been made possible due to the development of sensitive modern photomultipliers. Because of this, the existence of this radiation is no longer disputed, while its interpretation is still very much an open question.

Scientifically, this does not mean that the biophoton is any different from a normal photon, only that the way in which it is generated might be unique to biological systems. Though this far-reaching research question is often implicated in the usage of the term biophoton, most biologists have not yet seen the evidence that would justify such an implication (see below).

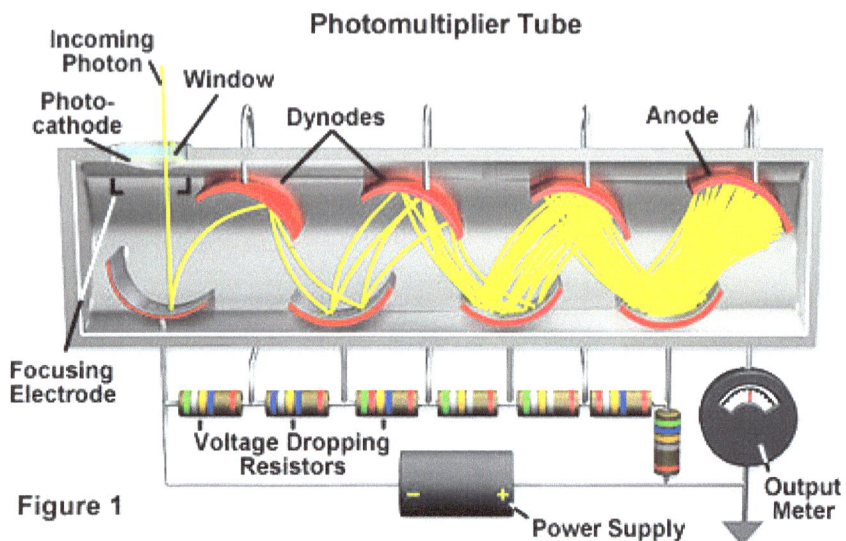

Figure 1

Studies have shown that injured cells will let off a higher photon rate than normal cells, and organisms with illnesses will likewise emit a brighter light, implying a sort of distress signal being given off, or healing energy may be at work in the injured cell. The biophoton communication system is faster and has a much wider bandwidth than electro-chemical nerve channels could ever achieve.

Researchers, F.A. Popp, Y Yan, A. Popp, E Humt, and S. Cohen discovered in Germany that:
- -The biophoton field of the human body is **almost fully coherent** – and as a consequence strongly coupled to all physiological functions.
-- It represents the regulatory activity not only from all the chemical reactivity in single cells but **performs the regulatory activity of the whole body**.
-- In this wholistic function it **displays all the biological rythms of the body**.
-- The measurement of the electrical parameters of the skin provide a powerful tool to look through the window of biological regulation because **skin resistance increases as photon emissions increase.**

COSMIC MECHANICS

-- Regulatory activities of the body are not stable functions of the electromagnetic fields within the body, but subjects of permanent, rythmical and oscillatory coherent field amplitudes.

-- The **sources of the photons are the DNA** and corresponding resonators in the cells.

-- The **photonic radiation is not the product but essentially the initiator of chemical reactions in the cells**. The radiation submits the information within and between cells.

-- The **photonic radiation is not limited to the optical range** but follows a frequency = constant - rule (the occupation probability of the phase space is equal for all wavelengths) and extends to longer wavelengths including the so called heat radiation of the body.

-- This photonic **radiation is the proper regulator and information carrier of life**.

The Marburg group of Fritz-Albert Popp calls this phenomenon **"biophotons"** in order to stress the difference to "bioluminescence": Biophotons are single quanta which are permanently and continuously emitted by all living systems. They are subjects of quantum physics and they display a universal phenomenon attributed to all living systems. World wide all scientists who agree with these statements call the radiation biophotons and the scientific field **"biophotonics"**.

The human body is an organization of cells. The cells must hold their positions and be stable terminals. They must do their job or the organism will die. The cells of the muscles, bones, eyes and ears must do their jobs in harmony for survival to occur. The cosmic being must handle or suppress confusions on the DNA level, align forces and flows and push action toward survival. Experimental results show that the health of a body is related to the coherence of the photonic communication between its cells. So we know that we must increase the coherence of the inter-cellular communication for healing to occur. How do we do that? You can just imagine the different ways that we could use to increase communication with the cells. We could talk to them, touch them, massage them or stroke them. You could even pray over them. It all works to increase the coherence level, and since every cell is in thorough communication with every other cell, when one improves, the coherence improves and all the other cells experience a greater coherence and well being. On the other hand, if you burn a few cells of your finger, the coherence of the photonic communication between all cells will decrease, resulting in the decrease of the well being of all cells in your body. Thus your emotional tone may drop from conservatism to antagonism when you burn you finger frying steaks at a back yard barbeque. Another example would be campers singing songs around a fire pit in the woods. One of the happy singers gets stung by a scorpion and her emotional tone descends to anger as she beats the pest to death with a log. The important lesson to be learned from this biophotonic discovery is that if we can raise the coherence of the biophotonic communication between the cells of the body of a sick person by any strategy or contrivance, healing will be greatly enhanced.

I've tried to get into thorough communication with a girl with a fever and the fever broke. I've tried to get a man to say "Hello" and "OK" to his sore throat and the soreness went away within 15 minutes. The biggest problem was to get the man over his consideration that it was silly to talk to his throat. If a person doesn't know cosmic mechanics he'll fall into agreement with scientific materialism and take toxic drugs instead of using the healing energy of zero-point and quantum waves.

One man had flu symptoms which completely disappeared in 30 minutes as I touched him, directed his attention, and got into live communication with him. This works like a charm every time with astonishing precision. It takes a little training to get a man on the street to perform miracles-as-usual with this technology because most humans cannot confront others and they are horribly out of communication with the human race.

When coherence decreases we have welts, cancer, sickness, viruses, germs and psychosomatic ills. When the orb is removed the coherence of inter-cellular biophotonic communication is vastly improved and these diseases become non-existent. Viruses and germs only feed on diseased cells. Diseased cells put out lots of biophotons which attract germs and viruses. Well being is achieved with a high degree of coherence.

In quantum physics, coherence means the subatomic waves are able to cooperate. These waves not only know about each other, but are interlinked by bands of common electromagnetic fields so that they can communicate together. They are like a multitude of tuning forks resonating in unison. As the waves get into phase or synch, they begin acting like one giant subatomic wave structure. It is difficult to tell them apart. Many of the strange quantum effects observed in a single wave apply to the whole. Something done to one of them will affect the others. Coherence establishes communication at a broadband level.

Popp wondered if we are eating the energy of photons when we take a bite of broccoli. Sunlight is the source of energy used during photosynthesis. When we eat plant foods we take in the photons and store them. When we digest it, it is metabolized into carbon dioxide (CO_2) and water plus the energy of the photons. The energy of these photons is distributed over the entire spectrum of electromagnetic frequencies and it becomes the driving force for all the molecules in our bodies. These scientists are treating the cells of our bodies like the employees of a business. To get a business into a high state of production you need good cooperation (coherence) between the manager, builder, maintenance man, shipper and accountant. The cells have most of the qualities of a person. They have a mind in the thought universe, gonads in the DNA, percetion and brain in the cell membrane, they eat and drink and eliminate, know their job, talk to the other workers (via photons) and hold down their position for the well-being of the organism (company). The well-being of the organism depends upon each cell doing its job just as the affluence of a company depends on each employee coordinating with the other staff and doing its job well.

Paul C. from Houston, Texas, had a Reiki master come over to teach him some things about chakras and spirits on 2/9/2005. Paul became very interested in his story and was amazed at his pictures he brought of spirits. He then was telling Paul about Chakras and light energy and

requested that Paul take a digital picture of him to see his aura. Paul was amazed to see the picture above. He was taken aside when Paul showed it to him and he said he has never ever taken a picture like it nor has anyone taken one of him like it before.(Note: The image was not tampered with and was taken straight from a Kodak dc240 zoom digicam.)

You'll notice (above) that every cell in the body is emitting light which penetrates clothing like infra-red rays do. Infra-red waves appear white in digital photographs. This is an example of intra-cellular communication of biophotons that have the same wavelength, are in phase, and carry coherent communications which keeps the organism in a state of peak fitness. The Reiki Master's viewpoint, in the center of the head, is radiating scalar waves that interact with space resonances in the air. Anyone in the presence of this Reiki Master will feel the effects of the order in his universe and will experience healing.

69) QUANTUM PROPERTIES OF LIGHT

Quantum processes dominate the fields of atomic and molecular physics. The treatment here is limited to a review of the characteristics of absorption, emission, and stimulated emission which are essential to an understanding of biophotons and lasers.

Atomic transitions which emit or absorb visible light are generally electronic transitions, which can be pictured in terms of electron jumps between quantized atomic energy levels.

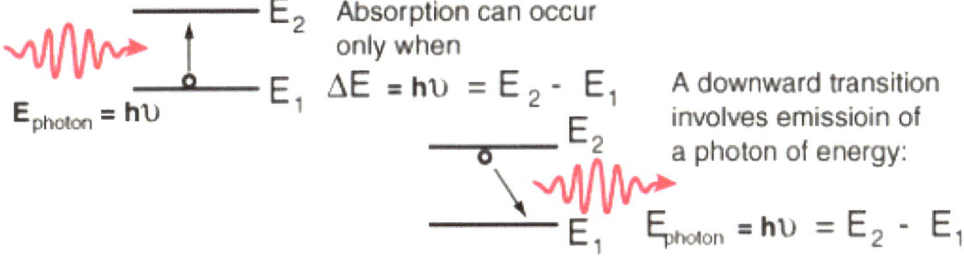

COHERENT LIGHT

Coherence is one of the unique properties of laser light. It arises from the stimulated emission process which provides the amplification. Since a common stimulus triggers the emission events which provide the amplified light, the emitted photons are "in step" and have a definite phase relation to each other. This coherence is described in terms of temporal coherence and spatial coherence, both of which are important in producing the interference which is used to produce holograms.

Ordinary light is not coherent because it comes from independent atoms which emit on time scales of about 10^-8 seconds. There is a degree of coherence in sources like the mercury green line and some other useful spectral sources, but their coherence does not approach that of a laser.

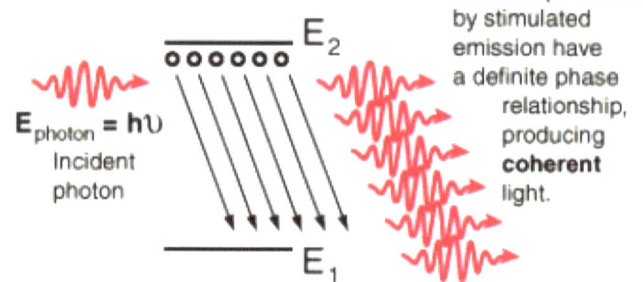

CHARACTERISTICS OF LASER LIGHT

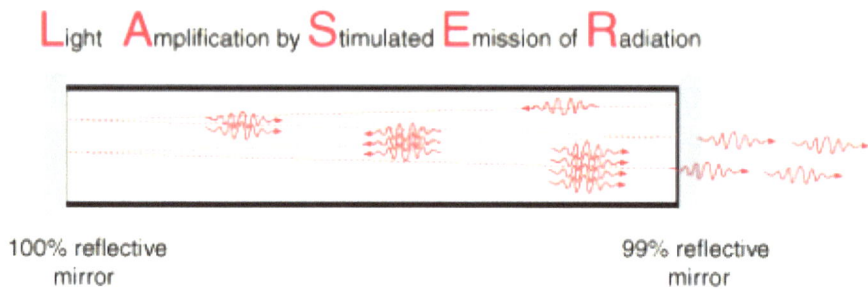

The stimulated emission of light is the crucial quantum process necessary for the operation of a laser.

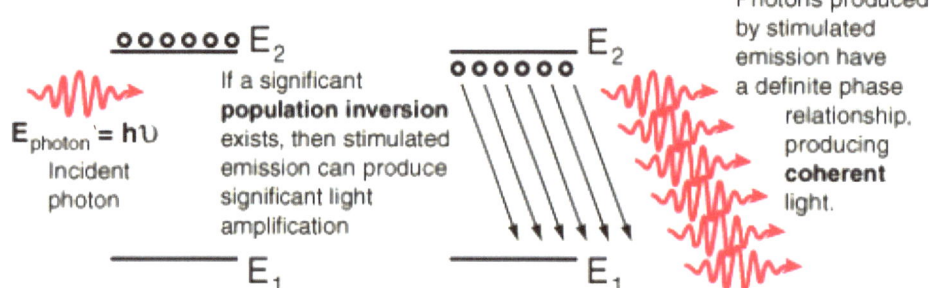

Fig. 1 The bad diagram. Light does not behave like this.

If figure 1 is wrong, then what is right? If we could see individual light waves, what would coherent light look like? Fortunately the explanation is quite simple. Take a look at figure 2A. That's what perfectly coherent light looks like. Scalar waves of subspace coming from a cosmic being can be represented this way too. Coherent light comes from very small light sources. A spirit or life-static produces photons from a view point of zero dimension. It is infinitely small. Coherent light is also called "sphere waves" or "plane waves."

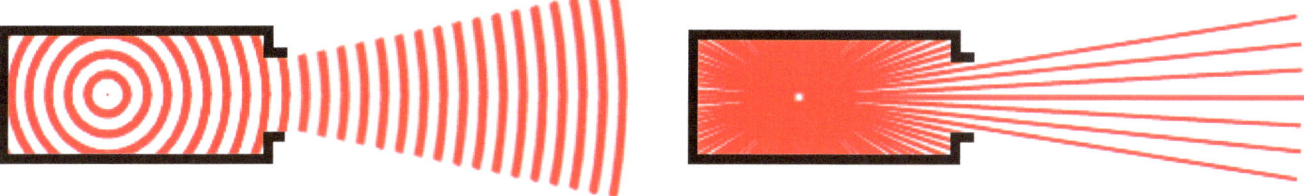

A.   **B.**

Fig. 2 A tiny light source emits waves. The emanations from a cosmic being come from a view point of zero dimension. Therefore, a cosmic being or life static is well qualified to produce coherent quantum waves of all frequencies.

A single small light source sends out electromagnetic waves in all directions as shown above. Of course this diagram is two-dimensional, while the real situation is 3D. Imagine a light wave to be spherical, like layers of an onion, but where the onion is expanding at the speed of light, and where the tiny light source is adding more layers in the middle. OR... we could imagine that the tiny light source is sending out a stream of particles in all directions. The paths of these particles are the "rays" of light.

OK, if spatially coherent light looks like an expanding bulls eye, then what does IN-coherent light look like? It looks like bunches of bulls eyes, or it looks like bunches of light rays where the rays cross each other. If we send our coherent light through a frosted screen, it becomes incoherent.

COHERENT LIGHT AND ITS EMMISSION IN LASERS

The coherent light produced by a laser differs from ordinary light in that it is made up of waves all of the same wavelength and all in phase (i.e., in step with each other); ordinary light contains many different wavelengths and phase relations. Both the laser and the maser find theoretical basis for their operation in the quantum theory. Electromagnetic radiation (e.g., light or microwaves) is emitted or absorbed by the atoms or molecules of a substance only at certain characteristic frequencies. According to the quantum theory, the electromagnetic energy is transmitted in discrete amounts (i.e., in units or packets) called quanta. A quantum of electromagnetic energy is called a photon. The energy carried by each photon is proportional to its frequency.

An atom or molecule of a substance usually does not emit energy; it is then said to be in a low-energy or ground state. When an atom or molecule in the ground state absorbs a photon, it is raised to a higher energy state, and is said to be excited. The substance spontaneously returns to a lower energy state by emitting a photon with a frequency proportional to the energy difference between the excited state and the lower state. In the simplest case, the substance will return directly to the ground state, emitting a single photon with the same frequency as the absorbed photon.

Soap Bubble Interference Colors

In a laser or maser, the atoms or molecules are excited so that more of them are at higher energy levels than are at lower energy levels, a condition known as an inverted population. The process of adding energy to produce an inverted population is called pumping. Once the atoms or molecules are in this excited state, they readily emit radiation. If a photon whose frequency corresponds to the energy difference between the excited state and the ground state strikes an excited atom, the atom is stimulated to emit a second photon of the same frequency, in phase with and in the same direction as the bombarding photon. The bombarding photon and the emitted photon may then each strike other excited atoms, stimulating further emissions of photons, all of the same frequency and all in phase. This produces a sudden burst of coherent radiation as all the atoms discharge in a rapid chain reaction. Often the laser is constructed so that the emitted light is reflected between opposite ends of a resonant cavity; an intense, highly focused light beam passes out through one end, which is only partially reflecting. If the atoms are pumped back to an excited state as soon as they are discharged, a steady beam of coherent light is produced.

INTERFERENCE

When a film of oil floating atop a body of water reflects light, a swirling mass of colors seems to magically appear. Yet, even though many people have seen such a sight, few realize that the cause of this strange phenomenon is interference between light waves. A simple soap bubble, like the one shown below, is another common example of interference, reflecting a variety of beautiful colors when illuminated by natural or artificial light sources.

This dynamic interplay of colors is generated by the simultaneous reflection of light from both the inside and outside surfaces of the bubble. The two surfaces are very close together, but the light reflected from the inner surface of the bubble must still travel further than light reflected from the outer surface. When the waves reflected from the inner and outer surface combine they interfere with each other, removing or reinforcing some parts of white light, resulting in the appearance of color. If the extra distance traveled by the inner light waves is exactly the wavelength of the outer light waves, then when the waves combine **constructive interference** occurs and bright colors of those wavelengths are produced. In places where the waves are out of step, **destructive interference** transpires, canceling the reflected light and the color.

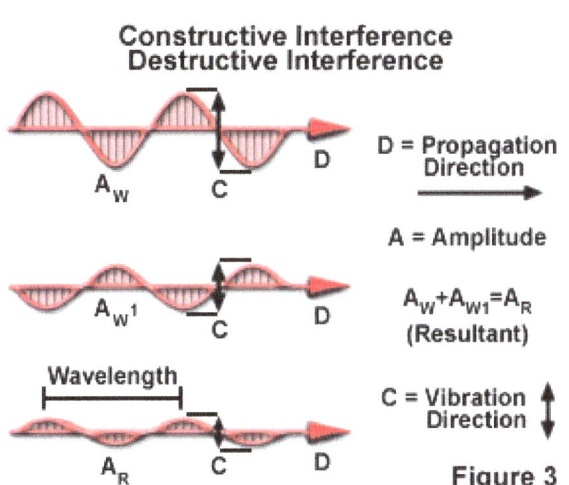

Figure 3

In order to better understand how light waves interfere with one another, consider a pair of light waves from the same source that are traveling in direction **D**, as illustrated below in Figure 2. If the vibrations, which are perpendicular to the propagation direction as represented by **C**, are parallel to each other and are also parallel with respect to the direction of vibration, then the light waves may interfere with each other. However, if the vibrations are not in the same plane or are vibrating at 90 degrees to each other, then they cannot interfere with one another.

Assuming all of the criteria listed above are met, then the waves can interfere either constructively or destructively with each other. If the crests of one wave coincide with the crests of the other, the amplitudes of the waves are additive. Thus, if the amplitudes of both waves are equal, the resultant amplitude is doubled. It is important to remember, however, that light intensity varies directly as the square of the amplitude. Thus, if the amplitude of a light wave is doubled, its intensity is quadrupled. Such additive interference is demonstrated in Figure 2 and is known as constructive interference.

Alternatively, if the crests of one wave coincide with the troughs of the other wave, the resultant amplitude is decreased, as illustrated in Figure 3. This destructive interference is accompanied by a decrease in light intensity and may even result in complete blackness if a total cancellation of light waves occurs.

70) EXPOSING THE DREAM WORLD WE BELIEVE TO BE REAL

The so-called 'mysteries' of life dissolve with every page. Who are we? What are we doing here? Who is controlling our reality? Why is there so much war, hunger, suffering and stress? What is 'mind' and 'emotion' and why do they control us? Who created religion and what are its advocates really worshipping? These questions and so many more are explained with a coherence and simplicity rarely, if ever, seen before. This book, supported by Quantum Physics, will change the reality – the 'life' – of everyone who has the courage to read it. The cutting edge just moved!
INFINITE UNDERSTANDING IS THE ONLY TRUTH, IT IS TOTAL AFFINITY, REALITY AND COMMUNICATION, EVERYTHING ELSE IS ILLUSION

IF IT IS STILL, IT'S TRUE, BECAUSE THE ULTIMATE TRUTH IS A STATIC (COSMIC BEING)

IF IT'S VIBRATING, IT'S ILLUSION, THE CREATION OF A COSMIC BEING

PARTICLES IN COLLISION IS THE ONLY COMMUNICATION THERE IS IN THE PHYSICAL UNIVERSE. There is no such thing as passive communication. For one being to communicate with another being they both have to put out anchor points and exchange energy.

71) WHAT DOES FAITH IN YOURSELF HAVE TO DO WITH REALITY?

I believe that most of the success and happiness that we experience in life comes as a direct result of our certainty regarding the nature of ourselves as a spiritual being. The pictures of orbs that we are getting is the smoking gun evidence that proves the existence of a spiritual world or "heaven." In fact, I hold it as an absolute truth for me personally that we are not human beings having an occasional spiritual experience such as going to church every Sunday or even sometimes once a year at Christmas. We are, in fact, spiritual beings learning to confront brief human experiences on this planet. So, from my point of view we are spiritual beings having a human experience.

COSMIC MECHANICS

This is all about our eternal quest. The word eternal means infinitely lasting, or forever without end. And the word quest means a special type of search. So, our eternal quest means a special type of search that goes to the ends of time.

When I was four years old I thought it strange that my Lutheran Sunday-School teachers pushed home the idea that Jesus was the Son of God and that he was omniscient, omnipresent and all-powerful. All this seemed suspect to me. No one had seen or talked to God or Jesus for two thousand years. I was four years old and I felt like I had lived forever. I wondered, "Why doesn't God ever seem to respond to my prayers?" I'd be given some type of a brush-off answer like, "That's how it is, Freddie Lee. Now stop asking questions and get back to church." As a good little altar boy I wasn't suppose to make too many waves.

Obviously this didn't satisfy me very much and I wanted more answers. As I got older and started to travel extensively throughout Europe and the Near East I became aware of and studied many of the world's major religions. Such as Islam, Judaism, Buddhism, Taoism, Catholicism, The Church of Jesus Christ of Latter Day Saints, Jehovah's Witnesses, Transcendental Meditation and many others. What I found was that all of us have a deep desire for spiritual awareness and freedom. We are all seekers. Considering our past as cosmic beings, this all makes sense to me.

Another anomaly I noticed as a child was a desire to fly around as if I were a ghost. Only now do I recognize that as a hold-over from the cosmic being that I was before I was born into this body. I still had the tendency to want to fly around after I took my new animal body. The problem was that animal bodies get damaged when you try to "fly" them out the window or take off like Superman.

Left, an orb visits the cemetery.

There are many different ways that people on this planet reach for spiritual freedom. What I also found was that most of them required a lot of faith and belief in their particular brand of the word of God. I wanted more than this and I want more than this for my friends. I wanted to know as an absolute fact that I was immortal, that I was not just a piece of meat; that I was not just an animal body and a brain; that I was in fact a cosmic being or soul controlling my body and brain. And therefore, when I died, that the body, brain and identity that everybody knew as Lee was going to be what was actually dying, but not me. I would never die. I was part of God. I was one of his viewpoints. Me, myself, as an immortal cosmic being, as a life-static, I'm going to go on forever as long as I don't get bogged down by my mental and material creations.

But the big question I had was the same question that billions of people down through the ages have had on this subject. "How am I going to know, as a fact, that I am a spiritual being?" It was a tough question to answer. But I think we've found a way to prove to ourselves that we are spiritual beings in human bodies. It is done by consciously getting out of the body. It is an Out of Body Experience.

So now I've got some questions for you. How is your life going to change if you know as an absolute fact, that you are an immortal, eternal, spiritual being? What would you be doing differently? How would you be spending your time? Would you stay at the same job? How old would you feel? How much more energy would you have? How would your goals change if you knew, not just believed, that you were eternal. Knowledge is certainty. Did you get that? So what we're going for here is nothing less than total certainty.

I am an eternal, immortal, unlimited cosmic being operating a physical, limited body. If I could step out of my body, I would know that I was outside my physical body and then I would know that I am a spiritual being and not my body. This is a differentiation. The essence of intelligence is the ability to differentiate. You perceive for yourself that you are not your body. You would get that reality really quick if this took place. You would see your body on the other side of the room and realize that you still had your memories. You are the same being in a slightly different state. It happens very quickly. And all of a sudden you'd say, "Well, for God's sake! Look at that!" The moment that hits you you're immortal! And you'd never be the same again.

That would be the ultimate proof for anyone. Beings placed under hypnosis do super-human feats. They feel no pain during operations and remember past lives. They press 50 pounds more iron than was possible in the normal waking state. They do Michael Jackson's moonwalk without ever practicing it and solve crimes by psychic abilities. People who develop their mental abilities are able to see into the future, astral travel as orbs, and do remote viewing. What all this is about is that we are something much, much more than just a physical body and no one has ever allowed us to discover our true potential.

Luiz Antonio Gasperetto is a painter who spiritually communicates with the ancient masters, VAN GOGH and other famous French Impressionist painters. He can paint a large picture in 6 minutes in the style of one of twelve famous artists from the past by channeling their energy. He can do it in absolute darkness with either his hands or his feet manipulating the brushes. Some claim you cannot tell the difference of the style to even their signature. How can he demonstrate such competence? He is in touch with his immortal nature. To the degree that you can be in touch with your immortal nature, is the degree to which you are in control of your life including emotions, physical body and relationships. The kicker is, you could do this too if you got rid of your orb.

What I have seen is real. I don't let anyone tell me that it's not. Nothing is true for me unless I myself have seen it. That disclaimer has to be balanced carefully with the warning against putting too many boundaries on my belief system which prevents me from looking in potentially rewarding areas. When you set no boundaries on your belief system, you impose no limits on knowledge. For example, long before Columbus the knowledge of the New World was limited by the belief that the Earth was flat. Mariners wouldn't entertain the concept that the Earth was round because "as every fool can plainly see, it's flat."

On the weekends some enlightened girls used to run a concession stand at Great America Amusement Park in San Jose guessing people's birthdates. They charged $5.00 per contestant and if they miss the year, month and day, the contestant wins a cute teddy bear. The crowds were flabbergasted at the precision of these girls. These girls know that they are cosmic beings operating animal bodies. They can withdraw from their bodies and look as a spiritual being in every direction from their viewpoint without the need of animal eyeballs. They can see and experience their own pictures in the Zero Point Field as well as every other person's orb. Nothing is hidden from them. Every mental image picture contains about 60 different senses, like touch, sight, sound, etc. The girls look at the mental image pictures of the contestants and follow them down the stack until they find pictures of the birth. Then they simply read off the date on the picture and astonish the contestant. There is no magic when you know as much as the magician. I was impressed when they did it to me. They could just as easily read off the date of my previous birth in my last incarnation. The kicker is; you could do this too if you got rid of your orb.

Of course most people who recover their super human abilities are helping others up the line. We create our own reality and control our own destiny and the way we do that is overwhelmingly through our intention. Intention is very much a spiritual attribute. When someone says to me that nothing ever happens in their lives and nothing ever seems to work and then, when I ask them what they really want to do they reply: "I don't know, really." Well if that is their state of being, their physical experience will reflect "I don't know really", and nothing of significance will happen. If, instead, you focus your intent on a specific goal, what you need to do to achieve it will always come toward you.

72) ENERGY MEDICINE

Everything in the Physical Universe, including our bodies, is composed of energy and information. So disease could then be re-defined as a disruption, distortion or interference arising within the information or energy fields within the body. A virus, or a parasite, or even a cancer, are just information fields. A most effective way to approach disease would be to correct these disturbances or change the information, which would eliminate the disharmonious energy creating the disease manifestation and, thereby, bring about healing. This is energy medicine. The conventional healing viewpoint that we have grown up with and that has dominated western medicine over the last 75 years is based upon biochemistry. (The science of chemical processes in the body and treating disease with laboratory produced substances) The energy viewpoint of healing deals with the electrical processes in the body and restoring its health and vitality by balancing and correcting them.

I believe that life-force energy is about as well understood today as gross electricity was 200 years ago. Practitioners of energy healing are working on a subatomic level. They are working with zero-point and quantum waves. In physics we know the observer affects the observed, and in its essence, energy medicine is a form of extremely highly concentrated non-intellectualized observation. Quantum waves are non-local and they communicate everywhere in the universe instantly. They are not limited like Physical Universe energy waves. Everything within its field, be it organic or non-organic, local or at a distance, will be profoundly affected.

The zero-point field itself has intelligence as does the system being treated, which directs how the healing will take place. I believe that this interaction of consciousness functions so far beyond the realms of human reality that man calls it miraculous. Clearly, life-force energy directly impacts the very physics and chemistry of matter. Water is changed by it; rocks hold healing energy, and beings can broadcast it to others. It is the reverse of voodoo. Life-force energy can be used for good or evil. If you aren't aware of it and aren't using it, some of your life-force is missing. Read *Quantum-Touch, The Power To Heal* (ISBN 1-55643-320-4) by Richard Gordon to learn to create life-force energy and use it to heal yourself and others. It is an essential human skill. I have no doubt that serious research into life-force energy will inevitably transform our conception of physics, chemistry, biology, medicine, psychology and botany leading to an entire new branch of science such as Quantum Healing. Including consciousness in science will not only change the sciences forever, but even more importantly, it will transform how we perceive our world and ourselves. It is a whole new paradigm. The impact of the awareness of the life-force energy will also occur on a very personal level. I believe that a few of the most likely changes will include:

- Healers would work in every hospital, emergency room, ambulance. By today's standards of patient recovery, the healing would look like something out of science fiction. The cutting, burning and drugs practiced by western medicine is very stone age, painful, degrading and expensive. Someone is needed to set bones and help with the delivery of babies, but chemotherapy, radiation treatment, cancer surgery, electroshock, trans-orbital leucotomy, and psychiatry are absolutely evil and destructive. They should have been banned last century.
- Hands-on healing would become a part of the education of every child from preschool on up. When a child falls, the other children would gather around to help. When a child is suffering from ADHD, he or she could quickly be brought into balance with energy work. By the time each child finishes their education, they would all be exceptionally powerful and wonderful practitioners.
- Professional sports teams would all embrace it.

- People would naturally and casually be doing healing work on each other in public. We will see them helping to remove headaches, relieving back pain or whatever, in line at movie theatres, cafes, bookstores, etc.
- People will be able to transcend religious dogma through their own spiritual presence.
- People will be able to express random acts of human kindness and love through healing.
- Unimaginable breakthroughs will be made.

When life-force is considered real, we will have a new paradigm through which to view the world. The way we grow food and what we eat will be evaluated in terms of how it affects our life-force. Education will be evaluated by how creative and loving processes enhance the child's life-force. Medical practices can be evaluated by how well the treatments enhance the patient's life-force, or biophotonic coherence. The value of exercise, yoga, pranayama, tai chi, and other various kinds of bodywork can take on a new importance. We can see how laughter, the honest expression of emotions, and the impact of love, care, tenderness, and touch enhance the life-force. When we consider the life-force real, we will live in a world that can transform its priorities and be a healthier and more fulfilling place for us all.

These Photoshop pictures are inserted to show how mechanical and stimulus – response the orb is. In 1/1,000 of a second the orb did the same dance each time the flash went off. I couldn't get permission to use the actual photos so I'm using these doctored images to show the effect. Orbs get these flinching reactions by being zapped in the distant past by some type of laser weapon.

COSMIC MECHANICS

Cranial Nerves: Distribution of Motor and Sensory Fibers

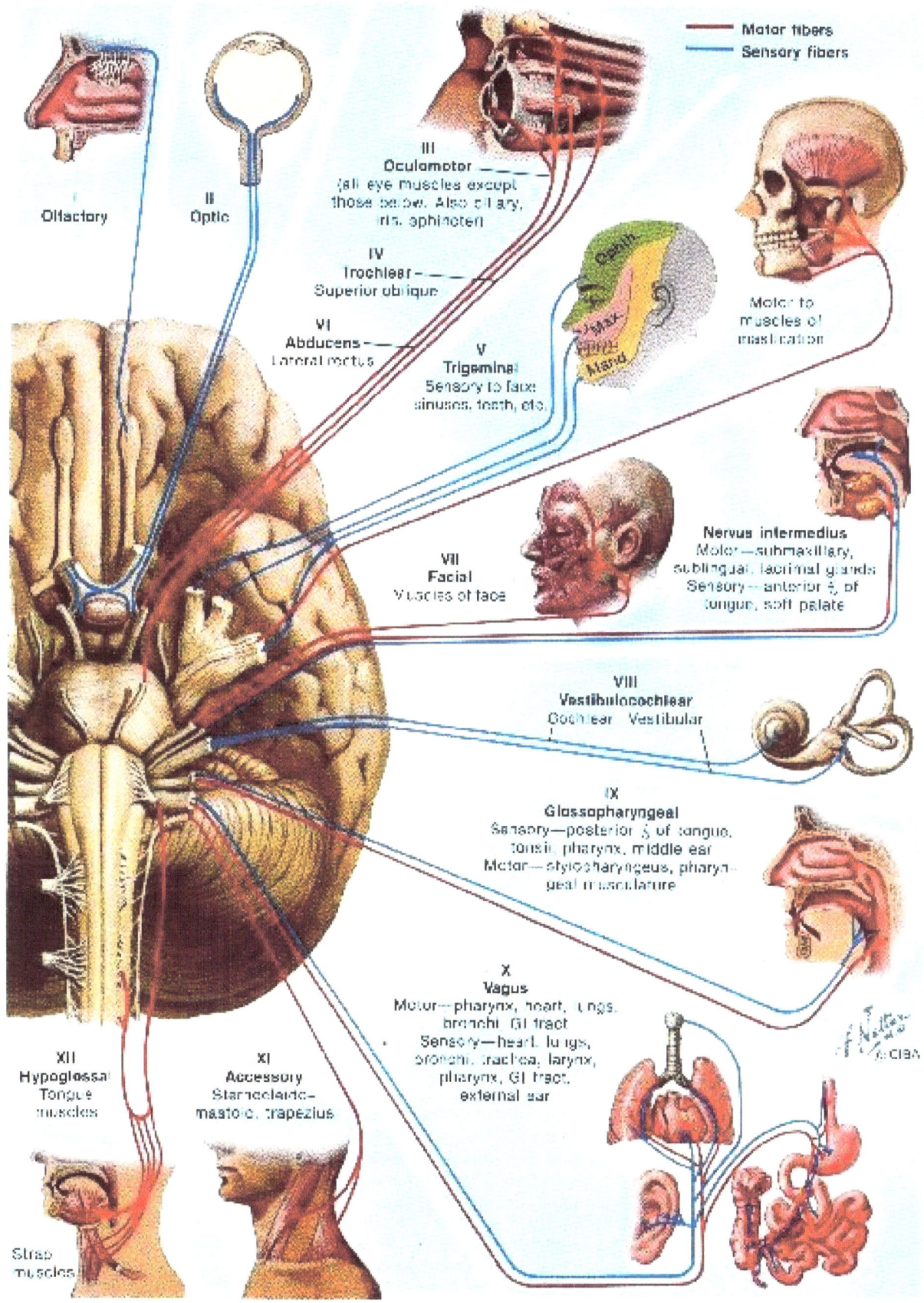

225

Just look at an anatomy book sometime to see how quantum mechanical the nerves and muscles are designed (above). It's absolutely fascinating! It reminds me of the wiring schematic of a Saturn 5 Rocket. The life force created this and don't think the life force cannot repair it after an accident.

Hands-On-Healing must be seen to be believed. By employing a very light touch you can profoundly accelerate the body's own healing response. You can actually see bones realigning themselves in response to the touch. The body knows absolutely what it needs to be strong and healthy. It only needs your support and help. Pain and inflammation are also quickly reduced and organs, systems and glands balance themselves. Each improvement increases the coherence of the biophotonic intercommunication or life-force energy. There is an immense joy and satisfaction in knowing that you can assist others in their healing; and this feeling is contagious. There is a connecting point between mind and matter where consciousness has an effect. It isn't on the macro level of muscle and bones, but on the micro level of quantum waves.

The DNA in each of our cells is so small that it is in direct communication with the biophotonic energy. That means that you can influence it directly with your mind; which creates quantum waves. You already know that your mind, not your brain, creates energy waves that control your body. That is how you walk and talk. You put out waves that make your muscles relax and contract. They push and pull on the body's skeletal structure causing your body to walk and talk. After a couple of years of practice as a baby, you grove in the quantum walking circuits that you create. So you walk easily.

Hands-On-Healing is an essential life skill. It will allow you to relieve an immense amount of pain and suffering from the lives of those you know and love. Wouldn't you like to live in a world where men can generously express their love and kindness by helping others to heal?

Don't think that you cannot do miraculous healings routinely. Do not limit yourself. In this life you've only been given inhibitions by your parents and teachers. You can work wonders! Let me show you. The ability to strikingly reduce pain and accelerate the healing process is an easily learned and invaluable skill. Whether you are a complete novice, a physician, chiropractor, acupuncturist, reflexologist, Qigong practitioner, massage therapist, or other health-care professional, Hands-On-Healing allows you a dimension of healing that heretofore has not seemed possible. Remarkably, this work amplifies the effectiveness of a wide spectrum of healing modalities.

In principal the Hands-On-Healing practitioner learns to focus and amplify life-force energy, which is most often referred to as "Chi" or "Prana". This is accomplished by combining various breathing and body awareness exercises. When the practitioner holds a high vibrational field of life-force energy around an affected area, through a process of resonance and entrainment, the client naturally matches the vibration of the practitioner, allowing one's own biological intelligence to do whatever healing it deems necessary. All healing is self-healing. So the greater the sickness and the shorter the time required by a person to heal himself defines the greatest healers.

There have been many studies made of the non-local effects of mental healing. How powerful is intention as a force and how infectious is the coherence of individual consciousness? Could it cure really serious diseases like cancer?

Elizabeth Targ, daughter of Stanford Research Institute remote viewer, Russel Targ, was a psychiatrist interested in alternative medicine. She was director of the California Pacific Medical Center Complimentary Research Institute. One of her studies showed that group therapy was as good as Prozac for treating depression in AIDS patients. She also read David Spiegel's work showing that group therapy dramatically increased the life expectancy for women with breast cancer. Her father's work in remote viewing strongly suggested the existence of some sort of extrasensory connection between people and a field that connected all things. She wondered if remote viewing could be used for something besides spying on the Soviets or predicting a horse race as she had once done. She knew that received wisdom was the enemy of good science and that the well designed experiment could even prove the miraculous.

She reviewed the experiments of biologist Dr. Bernard Grad of McGill University in Montreal. He was interested in determining whether psychic healers actually transmit energy to patients. Rather than using live humans, he used plants that he had soaked in salt water to retard growth. Before he soaked the seeds, however, he had a healer lay hands on one container of salt water which was to be used for one batch of seeds. The other container of salt water, which had not been exposed to the healer, would hold the remainder of the seeds. After the seeds were soaked in the two containers of salt water, the batch exposed to the water treated by the healer grew taller than the other batch. In a follow-up study, Grad had several psychiatric patients hold containers of ordinary water which were to be used again to sprout seeds. One patient, a man being treated for psychotic depression, was noticeably more depressed than the others. When Grad tried to sprout seeds in the water held by the depressed man, the growth was suppressed. This may explain why some people have green thumbs and others don't.

Grad moved on to mice that had been given skin wounds in the laboratory. He found that the skin of his test mice healed far more quickly when healers had treated them. Other studies have shown that people can influence yeast, fungi, and even isolated cancer cells. A biologist named Carroll Nash at St. Joseph's University in Philadelphia found that people could influence the growth rate of bacteria just by willing it so. Over 150 human trials had been done on healing at a distance. One showed that people could influence events just by 'hoping' everything would go well, even when they did not fully understand exactly what they were supposed to be hoping for. Elizabeth was concerned about sloppy protocol in the experiments. She teamed up with Fred Sicher to design the perfect experiment. It had to be double blind so that neither the patients nor the doctors knew who was being healed. All the healing was done remotely on a group of patients with the same conditions. The healers were from diverse backgrounds and covered the whole array of approaches. Each patient was treated by 10 different healers in turn who received no pay and no glory. Only a few healers were conventionally religious and carried out their work by praying to God or used rosary beads. There were several Christian healers, a handful of evangelicals, one Jewish kabalistic healer and a few Buddhists. Some were from the Barbara Brennan School of Light who worked with complex energy fields, attempting to change colors or vibrations in the patient's aura. Others used tones, visualizations, singing or ringing bells on behalf of the patient to reattune their chakras or energy centers. A few worked with crystals. A Lakota Sioux shaman used the Native American pipe ceremony. Drumming and chanting would enable him to go into a trance during which he would contact spirits on the patient's behalf. They also enlisted a Qigong master from China who sent harmonizing qi energy to the patients. The only criterion was that the healers believed that what they were using was going to work and that they had used it successfully earlier on hopeless cases. All information about each patient was kept in sealed envelopes. The only persons in the study who knew who was being healed were the healers themselves. During the six months of the trial period, 40 percent of the control group died, but all 10 of the patients in the healing group had become healthier. The treatment worked and it didn't matter what practice was used. A follow-up experiment proved that no matter what type of healing was used, no matter what their view of a higher being, the healers were dramatically contributing to the physical and psychological well-being of their patients.

These experiments didn't demonstrate whether God answers prayer or that he even exists. They demonstrated that when individuals outside of the hospital speak or think the first names of the patients with an attitude of prayer, they healed quicker and easier. It didn't seem to matter what healing modality is used so long as the intention for the patient to be healed was held. Calling on Spider Woman, a healing grandmother star figure common in the North American culture, was every bit as successful as calling on Jesus.

Several studies of heart patients have shown that isolation – from oneself, one's community and ones spirituality – rather than physical conditions, such as high cholesterol count, is one of the greatest contributors to disease. The people who live the longest are often not only those who believe in a higher spiritual being, but also those who have the strongest sense of belonging to a community. The intention of the healer is as important as the medicine. The frantic doctor who wishes his patient would cancel so he could have his lunch, the junior doctor who has stayed up for three nights straight, the doctor who doesn't like a particular patient – all may have a deleterious effect. **It might mean that the most important treatment any doctor can give is to hope for the health and well being of his or her patient.**

73) JOHN OF GOD -- HEALER

João Teixeira da Faria, commonly referred to as **John of God** or **João de Deus**, is arguably the most powerful unconscious medium alive today and is possibly the best-known healer of the past 2000 years. It is estimated that he has treated, either directly or indirectly, on the order of 15 million people during the past 40 years. João provides free treatment from his small hospital-style sanctuary in the town of Abadiania, Goias, Brazil - 130km from the capital Brasilia. The centre, known as Casa de Dom Inacio, is open three days each week; Wednesday, Thursday & Friday.

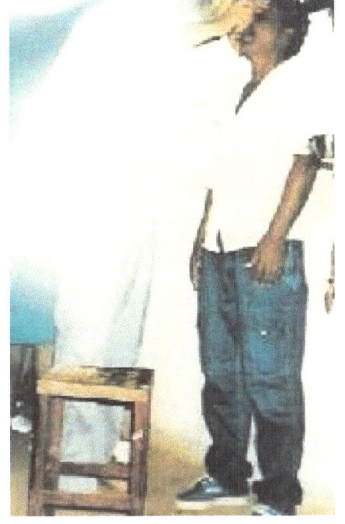
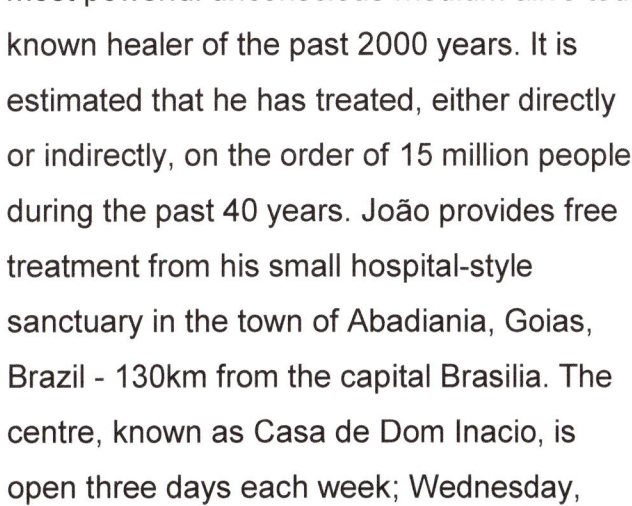

The medium João loses consciousness when he incorporates elevated spirit entities who then use his body to perform operations, treatments and cures of the physical and spiritual bodies. All who visit the centre may observe and participate in the proceedings. He will scrape away cataracts and eye tumors with a knife, remove breast cancers with a small incision and cause the crippled to walk with just the touch of his hand. In a meditation room a ceiling high stack of discarded crutches, wheelchairs and braces pays silent testimony to his success. He is acclaimed as the greatest healer of the past 2,000 years. The healing centre, affectionately known as The House, is simple, efficient and designed to cater to large numbers of people. The main hall is open at one end leading out to a covered walkway, toilets and a rose garden. This is the main treatment area where most of the visible operations are performed on patients where they stand. A lilting song heralds his entry. Followed by assistants carrying trays of surgical instruments, he moves slowly in trance. Anesthesia and sterilization are provided invisibly by the entities, a fact that astounds medical doctors who come to observe him. In 35 years there has never been a single case of septicemia in the house. Apart from the removal of eye tumors and cataracts, he also operates elsewhere in the body by scraping the eye. A small blue light emanates from the tip of his knife. This light strobes deep into the iris and seems to break up into multiple white fingers. At no time is there an anesthetic or aseptic used in any operation. Incisions produce virtually no bleeding, a fact which astounds doctors and scientists alike. Thousands of written testimonies are filed in the office along with piles of newspaper clippings and personal letters of gratitude. It is estimated he has cured more than 15 million people in his 38 years of healing. Reports by scientists from all over the world support claims of miraculous healings of every malady known and unknown, including incurable cancers, lupus disease and 139 cases (at the time of this writing) of AIDS. Many come with their white coats of skepticism, but all leave totally convinced that he is a genuine paranormal phenomenon of medicine. Like other renowned healers before him - Edgar Cayce, Ze Arigo and Britain's Harry Edwards, his miraculous cures provide hope for the terminally ill who have exhausted all avenues of modern medicine. Joao is the living proof of spirit existence that even the skeptics cannot refute. He also demonstrates that the only healing agency in the Universe is the spirit, and his photographs of healing show the wave structures that beings create which adds mass to this universe. The kicker is that you too can create these wonderful healing effects.

"A miracle is not the breaking of physical laws, but rather represents laws which are incomprehensible to us." – Guirdjieff

74) PRODUCTION OF EFFECTS

What is the cosmic mechanical operation of producing an effect? It is reaching and withdrawing, or pushing and pulling. A push and a pull by an immaterial cosmic being in the thought universe results in the creation of a zero-point wave. The cosmic being, or life static such as yourself, perturbs the infinite stillness to create these zero-point waves in the Thought Universe, which bump into each other and form space resonances that have the properties of mass, charge, spin and energy. This is the wave structure of matter. There is no discrete particle or solid chunk of anything in the whole Physical Universe, not even at the nucleus of heavy atoms.

When I was in high school my Physics text stated that atoms and molecules were mostly composed of space. I got the idea that an atom was like a miniature solar system with vast distances between the tiny orbiting particles. The text stated that if you took out all the space between the particles in the atoms of a car you would wind up with a chunk of matter no larger than a pin head with the weight of the car. Now I know that was false data because there aren't any particles at all. There isn't any solid substance in them at all and the direction of electron spin determines the charge, positive or negative. You can create trillions of zero-point waves with your imagination at once with no effort. The speed of reaching and withdrawing determines the frequency and wavelength. To put these into the Physical Universe requires standing waves resonating in space; space resonances. Every atom, electron and photon in the whole Physical Universe is in instantaneous communication with every other quantum "particle." Zero-point waves travel at superluminal velocities; they are not slowed down to light speed. All action in the Physical Universe is in extreme slow motion compared to the action of zero-point waves. Controlling the space resonances, which we experience as electrons and atoms, is difficult for humans, except for the atoms of our bodies which we control directly with our imagination by means of zero-point waves directing our cells. (See chapter on Biophotonics.)

These quantum wave structures built from the thought universe are part of you. **You are not distinct and unconnected from your environment.** The thought universe records the collisions of any wave structures the zero-point waves bump into. This is perception. **This is communication.** There is **no communication or perception without an energy exchange**. In the spirit world, or thought universe, all communication is done with images. In mortality we cannot hear, see, smell, taste or feel anything until the energy exchange brings us the data. You have the ability as a spirit to duplicate these collisions and remember them. The seat of memory is not the brain; it is one of your innate spiritual attributes as a cosmic being. You'll remember everything long after your body dies. Your present amnesia is not because you don't have any pictures of the past; it is the result of your inability to access those pictures. Under hypnosis you can access the pictures of earlier lifetimes because you are operating with the mindset of the hypnotist who is telling you that you can. When you awaken from the hypnotic trance, you cannot remember past lifetime pictures because your orb tells you that you can't. You feel like a small being only because your orb commands you to be small. Actually you are bigger than the whole Physical Universe.

As you will see, every electron and atom in the universe is in communication with every other part of it through zero-point and quantum energy interchanges. Our separation into discrete units as individuals is only part of the illusion of the Physical Universe. Things in our universe tend to clump together as in chemical bonding and gravity. **This is affinity.** It is produced by wave mechanics. When you pluck the A string of a guitar, the A string of another guitar close by begins to sing. **This is communication.** It is in sympathetic vibration. The other strings don't sound off. They don't share the affinity. Although the energy of the first A string travels through all the strings, only the A strings and their harmonics respond. **This is reality.** Everyone agrees that the A string is singing just as everyone agrees that the solid things of this universe are real. The solid things of our universe are resonating on a much more complex level than the simple string. The solid things are produced by the underlying sea of resonant zero-point waves producing standing waves of matter. These appear to be solid, heavy, durable chunks of mass but in truth they are simply interactions of zero-point waves. When large numbers of atoms clump together, their chaotic average motions can be predicted by Newtonian mathematics. You know, force equals mass times acceleration, distance equals half the acceleration multiplied by the time squared. Affinity, reality, and communication are the basic properties that we used to create our world. We could have used affinity, reality and communication to create a universe with different laws, but we used these ingredients in the recipe for the Physical Universe. And man, could we cook!

75) THE MECHANICS OF SPACE

Space is a viewpoint of dimension. That's all. It is a point to look from and distances between things. There is no space without a point to look from or without points to view. Space is not nothingness; it is caused by looking out from a point. It is how far you look to see things. So, we create our own space with our own viewpoints of dimension. Space only exists because we agree that we perceive through something that we call space. Space is created just by putting out points to view. Humans are trapped in the Physical Universe because they think that space has been created for them by the Physical Universe. They don't realize that they are creating all the space they will ever experience. Therefore, they will always be the effect of space and never discover the secret of their own divinity. If they knew that they could also conserve, alter or destroy space they would revolt and break their shackles. By retracting the points we put out to view, we can destroy the space we created. All spaces have the same properties (VIEWPOINT OF DIMENSION) except that they are defined by the particles contained in them. Space only exists relative to the objects we create. That also holds true for time and energy. Time and energy only exist relative to the objects we create. **Sound** is the compression and rarefaction of air particles which propagate in our atmosphere. **Heat** is the energetic collisions of particles in a mass. **Light** is the propagation of groups of electrons. **Radio** is the propagation of electrons from an antenna. **Electrical** energy is the movement of electrons through a solid, liquid or gaseous conductor. Some materials, like copper, readily conduct the spherical resonant standing wave electron energy from atom to atom. That's because they have polarized atoms which change position in response to the electric charge. But other materials, like glass, do not have polarized atoms and do not respond to that energy and act as insulators. **Magnetic** energy is the movement of electrons associated with electrical flows. Magnetic and Electrical forces always work together. You'll never find one without the other. **Gravity** is the affinity of all wave structures in the Physical Universe, which causes them to group together. **Inertia** is caused by the same quantum effects so that it is impossible to detect the difference between inertial and gravity effects.

If you search the world far and wide you'll never find any energy unrelated to a wave structure. That's because wave structures are the result of zero-point energy flows. Standing waves are not absorbed by the surroundings, while the rest are. And this is the secret to the construction of the Physical Universe where all particles were/are created as standing waves. Just like piano keys, at certain points in this frequency spectrum, scalar standing waves are allowed to develop and remain and are NOT absorbed by the surrounding wave cacophony. These, then, are all our permanent particles. Standing wave resonances exist in an infinite sea of zero-point waves.

76) KINESIOLOGY

Kinesiology is the study of the anatomy, physiology, and mechanics of body movement, especially in humans. A special branch of this science deals in assessing the body's state of health through muscle testing and energy flows. It is most commonly used to test food intolerance or food allergy by testing the muscles' resistance to force. It is a pain free technique which applies simple pressure to a group of muscles, while checking the body's resistance to the pressure.

A patient is given a sample of an herb or mineral to hold over his/her naval while the practitioner tests a patients muscle. If an improvement in the strength of the muscle is noted, then the patient is given that herb to build health. If strength is lost, the herb is discarded. It has been found that the herb itself does not have to be held by the patient. A vial of water with the vibrational energy of the herb will do for the test. If the herb test is positive, the patient does not have to eat the herb, drinking the water with the vibrational energy of the herb impressed upon it will suffice. There is also healing at a distance. If the vial of water that tested positive is placed on a sick person's picture, healing will take place no matter where in the universe the person is situated. This is possible because the zero-point waves that carry the life force energy to the cell communicate instantly with every other atom in the whole universe.

A little understood secret of kinesthesia is the effect of words on the chakras and biophotonic communications. A little folded piece of paper with the word "yes", written on it plainly will increase the muscle strength of a person holding it over their naval. Conversely, their strength will decrease if the note has "no" written on it. The spirit knows the significance of the writing and it will affect the coherence of the intercellular communication accordingly. Thus misunderstood words have a negative effect on the body's coherence and survival level. Never read on by a word that you do not understand completely, unless you hate yourself. The vibrational energy of each of the seven chakras can be put into crystals to strengthen not only the muscles, but also the emotions, intelligence and spiritual qualities of a patient. Again, as the coherence of the biophotonic communications increases, the health and well being of a patient increases.

It is hard for a person to feel the slight weakening of his muscles when he/she picks up the bit of paper with the word "no" written on it. But when he/she receives a complaint and summons from a process server, the effect is immediate; even before the summons is read. There is a whole scale of emotions and physiological reactions that humans experience as they reduce their health and well-being due to life's slings and arrows. A person could be enthusiastic about spending the day in Yosemite. The air is fresh, the sun is shining, the brook trout she had for breakfast is making her feel energetic and enthusiastic. Her intracellular biophotonic communication is at an extremely high coherence level.

When she returns to camp she notices that a bear has bent down the window of her car and is inside tearing hot dogs from plastic wrappers. This sight alarms her and reduces the coherence of her biophotonic communication instantly as she considers fighting the bear. She is now very conservative. She is angered by the damage the bear has done so she grabs her .22 caliber rifle and runs screaming at the bear to scare it away. The bear is cornered but manages to emerge from the window with a bag of French bread. The bear lunges at its attacker and the girl reacts in her rage by shooting the bear twice in the chest. The .22 shells don't stop the bear and the girl is overwhelmed by the shear size, strength, and swiftness of her enemy. Her intercellular biophotonic communication really starts to sizzle and the coherence drops as she realizes that her chances of survival are minimal. She swings the butt of her rifle at the bear who catches it and flips her and her rifle against a tree. She is dazed and apathetic as the bear bites the back of her neck and tears her head off. She dies shortly as the bear consumes the body and the struggle is over. The living cells are still communicating with photons but they know the end is near and lose the remainder of their coherence. The girl is now in anguish, out of her body, watching the bear rip it up and gulp it down. She is no longer putting out tractor beams and presser beams to control the body through its cells. She waits around to see what happens to her personal effects and car when her friends and park rangers discover the tragedy. Then she takes her orb, with the latest misadventure recorded in it, and looks for another body to continue her purposeless soap opera, *"As the World Churns."* Every succeeding life winds her up with more misadventures, more recorded pain and unconsciousness and an orb that overwhelms her.

77) HOW THE QUANTUM VACUUM COMMUNICATES INFORMATION

The Russian physicists G. I. Shipov, A. E. Akimov, and coworkers developed a "torsion-wave" theory that is enlightening. It shows how the vacuum can link physical events throughout space-time. According to the Russians, torsion waves link the universe at a group speed on the order of 10^9 c – one billion times the speed of light! It is standard knowledge that space resonances known as "particles" that have a quantum property known as "spin" also have a magnetic effect; they possess a specific magnetic momentum. The magnetic impulse is registered in the vacuum in the form of minute vortices. Like vortices in water, vacuum-based vortices have a nucleus around which circle other elements—H2O molecules in the case of water, virtual bosons (vacuum-based force wave structures) in the case of the zero-point field. These tiny vortices carry information in a manner analogous to the way magnetic impulses do on computer disks. The information carried by a given vortex corresponds to the magnetic momentum of the wave structure that created it; it is information on the state of that wave structure. As shown by Dr. Milo Wolff, these minute spinning structures travel through the vacuum, and they interact with each other. When two or more of these torsion waves meet, they form an interference pattern that integrates the strands of information on the wave structures that created them. This interference pattern carries information on the entire ensemble of wave structures from electrons to heavy atoms to living bodies to the stars.

In this way the vacuum records and carries information. Evidently there is no limit to the information that interfering vacuum torsion waves can conserve and convey. Ultimately, they carry information of the state of the whole universe. Throughout the universe, wave structures are linked by the vacuum in much the same way as objects are linked in the sea; by making and receiving waves. With the development of the Wave Structure of Matter by Dr. Milo Wolff we now have a model of energy and information transfer throughout the cosmos.

The wave structure of matter also explains the creation and perpetuation of our orbs. As the being experiences overwhelming pain and unconsciousness it creates a standing wave of energy complete with vortices which record the interference pattern of the incident. The interference pattern contains at least 60 different perceptics in the incident, not just sight, smell touch and taste. When the being, even trillions of years later, scans these vortices with its attention, the perceptics, pain and energy contained in them are delivered to the being and it re-experiences them. By completely duplicating those interference patterns the being can erase those patterns without a trace, get rid of the orb, and recover his abilities as a divine being. Of course, every cosmic being reading this who has an orb will object to it because their orb is continually denying them the fruits of observation, limiting what they can have, and insisting that they could never be a god. The only thing the orb ever gives you is inhibition. It is man's public enemy number one. Get rid of it!

How does the orb communicate with the person? We usually see orbs when they are out of the body, behind the head, at the feet or next to the person. How can they have such a diabolical influence on the person when they are not even touching them?

The answer to this question lies in the quantum wave qualities of the space resonances in the orb.. Experiments show that any two wave structures, be they electrons, neutrons, or photons, can originate at different points in space and in time; if they once come together within the same system of coordinates, that is enough for them to continue to act as part of the same quantum system even when they are separated. See Bell's Theorem above. When a large number of entangled wave structures are distributed through the orb, there is an instant transfer of information among them and with the cosmic being that created it without requiring that they be wired together or even be next to each other.

The physicist Nick Herbert said, "The essence of nonlocality is unmediated action-at-a-distance…A nonlocal interaction links up one location with another without crossing space, without decay, and without delay." The quantum theoretician, Henry Stapp said this linking could be the "most profound discovery in all of science."

At first blush, "action-at-a-distance" is strange. It is one of those things that you have to get used to in quantum physics. Einstein called it "spooky." But it is no stranger than many other aspects of the quantum domain. It is a validated and bonafide physical property that calls for a new paradigm in Science because the interaction involved in nonlocality is not any known form of interaction; it does not involve the expenditure of energy, and it transcends the hitherto known bounds of space and time. Nonlocal interaction is instant "informational" interaction, the fastest, most broadband communication system ever discovered.

78) THE GAME WE PLAY IS LET'S PRETEND

But we pretend we're not pretending. We decide to forget who we are but then forget that we've forgotten. We are the viewpoint that watches and evaluates and directs the show. We are the star in our own production and there are no extras. We are the I AM consciousness; that powerful, loving, perfect reflection of the cosmos, the cosmic being.

I hope that you are developing awareness in you of your unboundedness. You'll never look at life the same way again. You are a center of cosmic energy, a spirit trapped in a heavy, massive body. The only thing that could hold you back are your own considerations of your smallness, weakness and incompetence. You get that way by hanging out in a body for a few years. You think the Atlantic Ocean is so big because it takes so long to swim across it. You have no size. You can be smaller than a hydrogen atom or so large that you could put the complete Physical Universe in your vest pocket. You choose a size depending on what you want to do. To play ice hockey you should be about six feet tall and husky. You get the idea.

I like to think the unthinkable; to think outside the box. The information here has challenged me to do that. The measure of a cosmic being's recovery is its ability to accept the unacceptable, to disagree with what "everybody knows." You become overwhelmingly powerful as you disagree with the Physical Universe and make things go right through your awesome intention.

Could you possibly accept that far from being ordinary and powerless, you are extraordinary, eternal and all-powerful? This is unthinkable for most people given the current level of spiritual awareness in our society, but not for me. I am unlimited, powerful, capable and can and have caused things to go EXACTLY the way I wanted them to go in my life. That's really the true test of power; can you make things go right no matter what?

In your distant past you were an awesome being, powerful, knowing, and creative. You decided to reduce your knowingness so that you could play a game. You played games for trillions of years and have experienced the descending spiral. You had your own universe and were the god of it once. Now you are powerless and stuck in another universe. The **Secret of the Universe is this: you can get back your power and be totally free if you get rid of your orb**, that little black ball of mass that has suppressed you for two hundred trillion years. And the black magic spell is so deftly hidden that no one has come close to discovering it in many millions of years!

79) WHAT OUR ORBS DO TO US

So how could this be true when we all have amnesia and we've forgotten that we have it? The answer is that every super-conscious mind probed by hypno-therapists revealed a history of space travel, planet building, terra-forming and fantastic technology. Only our orbs forbid us to know. Our orbs are the source of our denial. Today, we have every ability and power that we've ever had. It is only our willingness to demonstrate those powers that has been denied us by our orbs. Under hypnosis we are dominated by the hypnotist and can perform miraculous feats, which our orbs would not allow. If so many people agree on this, it's real, even if our orbs have suppressed our memories and we cannot recall it consciously. Our past lives have been obliterated from our memories. We have no idea who we are so we assume that we are human bodies and we choose for ourselves human goals, not immortal goals. Don't try to change the world, but choose to change your mind about the world. This is what works. The goal is to become divinity! To grow and not look outside yourself for what can only be found inside. Don't be distracted by funerals you have to attend and life's emergencies. "Let the dead bury the dead," as Jesus would say. Nothing outside of you is going to save your spiritual bacon. Your orb is inside your own universe and nothing outside in the world is going to handle it. Not reading, not education, not the Pope, not Jesus, not God, not a Reiki Master; **nothing is going to handle your orb except you confronting it.** The biggest change you could make is the eradication of your orb. If you did that you wouldn't need education, you wouldn't need psycho-therapy, you wouldn't need psychic training, hypno-therapy or Jesus. That would make you totally free. Change YOUR mind... Change YOUR life.... The change has to come from within.

But getting rid of your orb isn't as easy as it may seem. You can't blast it with a Howitzer or burn it out with a laser weapon. It can withstand the Sun's fiery inferno. It is an insidious little demon built over two hundred trillion years with irresponsible postulates and the most horrific conflicts of purposes. You cannot trust your feelings regarding it. In the Star Wars story Luke Skywalker is taught by Yoda to trust his feelings while wielding his weapon and fighting the dark side. But you cannot differentiate the source of your feelings. Maybe it is a cell demanding a Coke; maybe its your body's zero-point entity demanding sex; but usually its your orb demanding that you follow its orders or it will make you sick. Most of your feelings come from the orb, so you cannot trust those. The orb will put up a hell of a fight to survive as itself, while it caves you in. If you try to take control of it and use shear will-power to confront something for two hours, in spite of it's telling you not to, it will make you sick. If you try to use your will power to do the right thing instead of following its command to strike back at the person who offended you, it will make you sick and miserable. If you try to get others who have orbs to help you get rid of yours, they will strike out at you. If you try to help others vanquish their orbs, they will resist your efforts. It works like in the movie, *Matrix*, the very people you are trying to help will turn on you.

Let's take a look at the Achilles' heel of an orb. Does it have any week defenses? Yes, but power isn't one of them. It has the power to put butterflies in your stomach, turn on bizarre aches and pains all over your body, make you deathly sick and make you unconscious if you don't capitulate to its commands. It is the ultimate terrorist, and you are in bed with the enemy. When you understand your position you will find that you are locked in mortal combat with an evil horde of faceless warriors holding you at gunpoint. If you fight these hordes with anger, revenge and hatred you are finished. An angry person will never tell the truth. You've probably noticed that, and you need the ultimate truth to handle your orb. You have to fight this awesome enemy as if you were god, which in fact you are. You have to fight with powerful intention, nobility, enthusiasm, determination, responsibility and steely-eyed joy. You cannot let it get you down. That's why you'll need a friend to help you wage the battle. The two of you could be more powerful than the orb, because the orb is not holding a proverbial gun to your friend's temple, just yours. The enemy orb can be identified with an absolutely negative force and overwhelming power but very little intelligence. **It is a total machine which has no consciousness and no way to change its strategies. It can be undone with intelligent strategy or technology. That is its Achille's heal.** The way to oppose it is to use fantasy, invention, cleverness and creativity. You can consider the enemy orb a rigid and unavoidable opposition that can be contrasted only with elasticity and fantasy. If you get the idea of Robin Hood and his merry men, that should help. The Sheriff of Sherwood Forest has the manpower, treasury, weapons, law and King on his side. But he and his men are short on intelligence and leadership as they perfunctorily carry out their police duties. So Robin's band uses clever shifts, ambushes, disguises and impersonations to make a fool of the Sheriff and relieve him of his ill begotten riches and redistribute it to the poor. They do all this with an enthusiasm and esprit de corps that is inspiring. The orb must be handled with a similar enthusiasm.

Humans see life and reality through the limiting perspective of a microscope. They make mountains out of mole hills while ignoring the unifying, harmonizing macroscopic realities that lie just beyond their limited views. It's like they are wearing blinders. Blinders are used on horses to keep them from being frightened by what they would see if they had broader vision. People do the same thing. They pull the blinders in tight to block out or condemn things that are different from what they are used to. The orb seeks to knock out the good and perpetuate the bad. This leaves them with a very limited but comfortable microscopic view of reality instead of a limitless and challenging macroscopic view of themselves, others, and their relationship to the universe. This is what happened to Galileo Galilei. His contemporaries had a limited but comfortable view of reality and Galileo's assertions made them very uncomfortable.

The Technology Required to Get Rid of an Orb

There is only one way to rehabilitate yourself. You have to get rid of your orb. An electro-psycho meter can help you locate the little ridges of charge in your orb and live communication will blow them. You'll need some help. When the ridges are all gone you'll have your memory back, you'll be able to create energy and your life will be "miracles as usual" again.

Extremely powerful methods have been developed to eradicate orbs. Carlos Castaneda's technique of recapitulation was one of the first techniques developed to annihilate standing waves in the cosmic being's universe. Since these standing waves are quantum entangled with the space resonances of the orb in the material universe, these are also annihilated simultaneously. The purpose of Castaneda's technique is to release energy knots relating to past experiences or interactions. The method is to move backwards, re-experiencing the past occurrences and releasing the energy that has remained stuck with them. Recapitulation frees the energy entangled in painful memories, thereby deleting the ridges that comprise the orb. Carlos Castaneda's methods have fallen out of use because they could not be taught successfully to others.

L. Ron Hubbard developed the technology of Dianetics to rid people of their reactive minds. This is what the History Channel has to say about L. Ron Hubbard, *This Day In History*, **May 9, 1950:**

On this day in 1950, Dianetics, The Modern Science of Mental Health, was published. Former science fiction writer, L. Ron Hubbard, wrote that deep unconscious memories are the source of all unhappiness, and that he had the key to erasing them. Participants work with an auditor, a counselor trained to uncover and delete unconscious memories. An electro-psycho meter aids the auditor in locating areas of spiritual distress. By 1954 Hubbard's methods and philosophy had evolved into the Church of Scientology, which claims to lead to spiritual release and freedom if practiced correctly. Millions of copies of Dianetics have been sold and Scientology continues to grow. And that's how this day went down in history, May 9th, 1950.

80) THE SECRET OF THE UNIVERSE

There has to be some fundamental truth underlying reality as we know it for UFOs, alien abductions, hypnotism, mysticism, spiritualism, out of body travels, telekinesis, mental telepathy, and near death experiences to exist. Have you seen the paranormal and New Age material on the internet? It goes on for thousands of pages. A new paradigm must be found which explains all these phenomena. The Elite who control this planet are looking for this eternal truth also. They haven't discovered it yet despite their access to unlimited funds and crashed disks from Roswell, Corona and other sites. The secret must be so arcane and sophisticated that no one will ever figure it out. Or it must be so obvious and "in your face" that it has been overlooked for centuries. The latter is the truth of the matter.

The theory which works is, the ultimate truth is a zero! Its ultimate simplicity is a little difficult to grasp at first. But it is what everyone, including the universities and the government has been missing for centuries. You say. "That's just a theory and you can buy theories for a dime a dozen!" Yes, you're right. I'd be just as happy if this theory went away or was burned up and forgotten. It would be more endearing to believe in the theory that this universe was created by a God of love. That love and sex make the world go round. Sigmund Freud tried that one but it wasn't workable. It didn't fit all the phenomena.

COSMIC MECHANICS

Life is basically composed of two things, the Physical Universe and an X-factor. The X-Factor is the thing which organizes and mobilizes the Physical Universe. The Elite would love to discover how it operates, and it is the easiest problem anybody ever confronted. It is idiotically simple. Anybody with the common sense that God gave gravel could have solved it. That's why it was always a mystery. No one had ever thought something that simple could do so much. The X-Factor obviously has no wavelength; it has no energy in it so it couldn't have any space or time. It is a zero! A nothingness! It's a simple dimensionless point from which to create everything that exists. It is a cosmic being! It is us!

We have a big Homo sapiens that we take with us everywhere we go. It seems so strange! Our Homo sapiens opens its eyes every morning and we look out through them. We brush its teeth, comb its hair and walk it out the door. It doesn't even belong to us. We stole it from its parents when it was a baby, and we'll relinquish it to the Physical Universe when it finally falls over, dead. The stuff it is made of doesn't even belong to our own universe. That's why we obsessively create these orbs. The two universes come crashing together with more force and pain than we can confront. Our universes get thoroughly crushed into the Physical Universe during times of great pain and unconsciousness. This causes quantum entanglement of wave structures and little ridges to form on our orbs that last forever, just like magnetism lasts forever. The little ridges have the interference pattern pictures of the event that created them. Cosmic beings have the ability to duplicate the events that they experience, but the events recorded on the ridges are solidly recorded on space resonances as physical impressions. DVDs, CDs, films and phonograph records are similarly recorded on physical media. That's why these recordings are so damaging. It wouldn't hurt a cosmic being to remember a bad experience such as being burned at the stake. But when it makes an interference pattern recording in the matter, energy, space and time of the orb, whenever the being's attention scans it, it delivers the force, pain, burning sensation and unconsciousness that the being experienced at the time of the actual burning! That's why these orbs are so overwhelming and so hard to handle. A cosmic being will look anywhere except at the orb because it knows that act would be fatal!

TIME: Time is havingness. It is energy and structure in relation to space. Man creates things in this Universe by being something which acts to produce a product. It is BE, DO, HAVE. The cosmic beings are devoid of havingness. They just exist with nothing. Then they put out attention units which have wavelength in the stillness which causes space resonances (quantum particles or electrons). If lots of electrons are created by lots of cosmic beings and these electrons are compressed within giant stars by gravity, heavy matter is produced in a supernova. Matter is condensed wave energy. The frequency of the electron results in time, which is the havingness of the Physical Universe. As time changes, the things we have change.

So the cosmic beings creation results in an understanding, the wonderful random Physical Universe that we all play in and experience our adventures in. It gives us our bodies, vehicles, homes, cities, planets, solar systems, galaxies and the rest of the Universe. A universe is a whole system of created things. We have our own universes, the whole system of created things that we have produced with our own imaginations. Our own universes should be trillions of times brighter and more forceful that the Physical Universe, but alas, we have given up our own power and force. Now we depend for energy upon the Physical Universe. We use the Physical Universe energy: food, fuel, electricity, hydraulics and sunlight to place things in space and time instead of our own energy.

When a cosmic being gets degraded and loses its power it seeks a smaller, weaker game and attaches itself to a human or an animal. It has lost the ability to generate force itself, so it utilizes the force of the animal body to confront the conditions of the Physical Universe. A cosmic being figures that any game is better than no game. When the debilitated being becomes a human it plays the same games that it had played for trillions of years. As above, so below. It cares for the body, creates randomity within social situations, gets "help" from the psychs, watches picture shows and goes to war over trivialities. A patriot is a person who will kill himself and others over trivialities.

Mortal Christians conceive heaven to be a place of total peace and pleasure. But that would be a hell for you. If you didn't have your orb you would experience a perfect balance of all opposites, a totally accepting experience of all pain and all pleasure, all hate and all love, all ugliness and all beauty, all fear and all conflict, with all calmness and all peace. In other words, your unbounded self, the part which is divine, the zero, exhilarates in the ultimate excitement, enjoyment, variety, creativity, --which is truly heaven. You might believe that's only for God to experience. Only when you have evolved to the awareness that you are the unbounded creativeness, the zero, created in the image of God, will you be able to experience this acceptance of all that is, all that ever was, and all that ever will be, as perfect.

Scientific remote viewers like Courtney Brown (Author of *Cosmic Voyage*) describe a cosmic beings experience in the thought universe. They feel a slight tugging motion from some unseen force as they drift. The distances seemed unlimited. There is endless space, it goes on forever. They are dazzled by the eternal world spread out before them and it seems like somewhere within lies the nucleus of creation. When they look at the fully opened canopy around them they see none of the inky blackness we associate with deep space. There are various degrees of luminescence. Orbs drift into the 'Grand Arena' or staging area where the orbs of departing humans are accumulated and dispersed to their ultimate destination. This area is similar to the hub of a great wagon wheel where millions of returning souls are conveyed in a spiritual form of mass transit from the center along the spokes to their proper destinations. These souls look like myriads of orbs, sharp star lights going different directions. Some move fast while others drift. The more distant energy concentrations look like islands of misty veils. The most outstanding characteristic of the thought universe is a continuous feeling of a powerful mental force directing everything in wonderful harmony. This is a place of pure thought. These are images of motion. Orbs are swept along by a strong current and they join thousands of orbs moving toward a distant light. The Earth is just a speck in the distance. There is such a wonderful peace that no one wants to return to the physical. As they move along they pass patches of lights in clumps separated by galleries. The galleries are like long corridors bulging out in places containing millions of orbs. The shining orbs are the souls of people within the bulging galleries reflecting light outward. That's what you might see between lives, patches of lights bobbing around. Here no one is a stranger. There is absolutely no hostility toward anyone. That's because spirits recognize a universal bond which makes them all the same. They truly are created in God's image for he duplicated himself innumerable times to fill all the universes. All of God's children have his capabilities and potentials. They are all zeros without matter, form and void. They are a perfect nothingness that can create without limit matter, energy, space and time. There is no suspicion toward each other. There is nothing to hide. One look at a being and you know everything about it. There is no sex or gender with spirits but some mock themselves up with masculine or feminine traits to demonstrate their preferences. There is complete openness and acceptance.

Am I saying that to a cosmic being everything is perfect? Things like domestic violence, murder, rape, mayhem, disease, death and selfish little mortals? No, but they are all games and to a degraded being any game is better than no game. Perfect responsibility is the willingness to assume that you are the creator of all efforts and counter-efforts in every universe, your universe, the other guy's universe and the Physical Universe. Taking full responsibility doesn't mean taking the blame for the mess mankind is in. Fault is entirely different. Finding fault will only beef up the orb and worsen inter-personal relations. You take full responsibility as soon as you recognize that you are the cause of every effect you have ever experienced, and not a moment earlier. Reducing your sense of responsibility is the main thrust of the insidious orb. It denies you in every conceivable way. It tells you that you are small, insignificant, powerless, hopeless, stupid, careless, unconscious and untrustworthy 24 hours a day, 7 days a week. It never quits. And you've been duped into thinking that those are your own thoughts! Don't listen to it! All beings have the potential of knowing everything. It is only the barriers which each of us have created for ourselves which prevent us from knowing completely and with certainty. No one could possibly damage us; after all, we are a nothing! Only those things that are intimate to us can be damaged – like our bodies, families, pets, houses, cars and possessions. So only our own decisions to be small and stupid could possibly affect us. Get with a counselor and rid yourself of those self defeating attitudes and considerations.

This is a two terminal universe. You can't have pleasure without pain. You can't have a god without a devil. All the positives and negatives cancel each other out. All the electrons and positrons cancel each other out. All the matter and anti-matter sum to zero. But, of course, it is impossible to experience this at the mortal level when we are plagued with orbs. That's what the mystics meant when they said that all is illusion or Maya. But it's such a real illusion to us Homo sapiens! Of course, who could enjoy an exciting play without being able to temporarily forget that he was just watching actors upon a stage playing parts written by an unseen author for entertainment purposes? I take Shakespeare's view that the entire world is a stage and all the men and women only players who, in their time, play many parts. A good actor temporarily loses himself in his part. It is the same with a man. He has temporarily lost himself in a part and has forgotten that he is the only being who choose it. Each soul has chosen every part it plays so there is no injustice from the larger perspective.

COSMIC MECHANICS

Some spiritual advisors tell us that with each incarnation we learn valuable lessons that will eventually allow us to transcend the physical, reach enlightenment, or attain Nirvana. Those teachers did not know about the 21 gram orb that has been degrading us for 200 trillion years. It matters not how many lessons we learn, Christs we accept as our personal savior or pain and agony that we overcome. No amount of study and learning will deliver us from the clutches of our orbs. But ten minutes in the hands of a master who can direct our attention to our orb's pictures can knock a hole in the orb. It will bring our awareness from <u>not-know</u> up to <u>know</u>. It will free us from the accidents to our right shoulder that we have been experiencing for the last two million years. Or it will eject us from the top of our heads so we can look around and realize that we are not our body, and never have been.

This is my book on Cosmic Mechanics. Such strange ideas! But I have a right to these ideas. This is how far I've come in my understanding. But I have not arrived at the ultimate truth. I must not let my misunderstandings bar the route out for myself and others. I am not an authority. I may be wrong. These are my ideas and you may not agree with them. What is true for you is what agrees with your own experience. I'm not a Guru. I'm a searcher like you. But I humbly present these viewpoints for your evaluation. You may be able to improve on what I've done. If so, I'd like to hear from you. We all have this brief window of time to improve our condition. If we work vigorously, our efforts will bring us at last to the vast unlimited vision of freedom and truth and the glory of being really us. Do not fear that we are inadequate. It is better to be afraid that we are powerful beyond measure; to be very afraid of the consequences of our power! Ask yourself, who am I to be brilliant, gorgeous, talented, fabulous? Actually, who are you not to be? You can be anything you can conceive! You are a cosmic being, a nothingness, a matter, energy, space, time production unit. You were the first laser weapon. You are a Creator, one of God's viewpoints. As such, we have the power to overcome our reptilian "custodians."

Until now, no one has ever permitted us to know how great we really are. Our parents, siblings, teachers, police, governors, presidents, kings, emperors and secret societies taught us that we were fragile human beings and they told us to be careful. Psychologists told us that great expectations lead to great disappointments. The only thing any of them ever granted us was inhibition. As children we were told to be careful, eat our vitamins, wear our rubbers, don't take candy from strangers and don't play in mud puddles. We had a constant tirade of what we're supposed to do or not do with our bodies. But nobody ever told a kid "Your ability to handle everything you see depends on your willingness to tolerate the actions of others or manage them at will, depends upon your ability to show mercy when in a position of trust, depends on your ability to be the captain of that ship, your body, The only insurance there is, is your ability and willingness. By not telling him that, we have predicted a life for him of turmoil, confusion and sickness. If the cosmic being is in control of the organism he can expect a healthy body and a successful life. But if a machine, an orb, is in control of our bodies, we have sickness, depression, a hell on earth.

Our playing small does not serve the world. There is nothing enlightening about shrinking so that we won't harm others or cause them to fear us. Think big thoughts, positive thoughts and make a future for yourself. As you let your light shine, you unconsciously give other people permission to do the same. As you are liberated from your own fear, your presence automatically liberates others.

81) THE MECHANICS OF SPACE RESONANCE

I was lacking the knowledge of the cause and structure of the "particles" that make up the Physical Universe. I knew it had to be closely connected with quantum physics. I knew that it was all waves, but what causes waves to bunch up to create gravity, size, time and charge?

Along came **Dr. Milo Wolff (left) to the rescue. He corrected the mathematics of the Schrödinger wave equation and discovered an intuitive theory that demonstrated that energy and information transfers from spiritual beings to quantum waves to physical reality.** Dr. Wolff has given us a synthesis of science and spirit, a reconciliation of these two domains of wisdom. He also developed the mathematics to aid in the application of the knowledge.

Now, at last, I could understand where this universe came from. He connected the spiritual cause with the material effect in an intuitive way. The following technology of space resonance and the Wave Structure of Matter (WSM) is based on the discoveries and writings of Dr. Milo Wolff and are used with his permission.

COSMIC MECHANICS

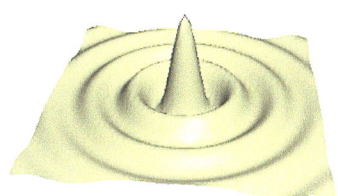

This illustration of the Space Resonance wave of an electron lets you imagine vividly what an electron looks like if it were big enough to see. The wave intensity is shown in two dimensions. In space, the resonant waves exist in three dimensions so the illustration would look more like an onion with concentric crests and troughs extending from the center outward.

The physical structure of the electron involves inward and outward spherical quantum waves. At the center, spherical rotation of $720°$ changes in-waves to out-waves. We are referring to quantum scalar (zero-point) waves arising out of the infinite stillness of the thought universe to create the quantum field. The stillness can be agitated or perturbed by spiritual beings. These perturbations travel instantly to the ends of the universe as zero-point waves which are invisible and undetectable by physical devices but which can be intercepted by spiritual beings like us. We are part spiritual and part physical in a body. As spiritual beings we can communicate with zero-point waves.

Any difference of potential resisted, creates energy. The static resisting motion creates energy. Effort moving matter produces energy. The effort of a magnetic field resisting the motion of a copper wire produces electrical energy. A cosmic being is an electrical field more capable of high potential and types of waves than are known to nuclear physicists. As physical beings we can communicate with physical wave structures. We are the interface between two universes, our own universe and the Physical Universe. Cosmic beings create energy and then condense the space which contains the energy – which results in matter. Matter gets twisted together in standing waves and then persists as a nonlinearity over time as expressed in quantum mechanics. That's why each electron has a spin and each gallaxy has a spin. There is a $720°$ spin in each space resonance. This is the basis for the spiritual creation of the universe and why material things retain frequencies and memories and have morphogenic qualities. The Physical Universe is more thought than things. It comes from the thought universe. It is Maya, illusion, imagination. It is solid because we consider it to be so.

The simplest resonance is the electron whose mathematical physical structure is exactly known. Electron waves (comprising all charged particles) extend to infinity serving as the 'communicator' of the natural laws. The entire universe operates at the frequency of the electron. It is the beat of the Universe. Everything in the entire universe is synchronised to it.

Dr. F. Lee Aeilts

QUANTUM THEORY ENIGMAS

Four thousand years ago, Democritis created the point particle of mass to represent the fundamental elements of matter. This concept was satisfactory until about 1900 when quantum properties of matter were found. Then, puzzles, problems, and paradoxes appeared because most properties of matter derive from the wave structure of particles. Democritis couldn't know this and until recently few persons challenged his embedded concept. Nevertheless, Schroedinger, deBroglie, Dirac, and Einstein, the founders of quantum theory, preferred a wave structure of matter, and in the last decades researchers have validated their intuition.

The Space Resonance concept – matter structured of spherical wave centers – avoids and explains the paradoxes and problems of point particles. In hindsight it is simple; since mass and charge substances do not exist in nature, removing them from particle structure also removes their problems. In their place, the wave centers possess the properties of mass and charge which we observe in a human-sized laboratory, but without the problems of finding mass points which do not exist!. One of the fascinating puzzles explained below by this new structure is the former mystery—the spin of the electron! The overwhelming proof of the Wave Structure of Matter is the discovery that all the former empirical natural laws originate from the wave structure. The probability of a coincidence is infinitesimally small.

New Insights A simple spherical wave structure of the particles leads to new exciting insights, including: 1) the origin of the natural laws, and 2) the relationships between the smallest things—particles—and the largest, the universe itself. These insights are breath-taking in their scope and potential. This structure appears to agree with and predict experimental observations.

Origin of the Laws

The origin of the natual laws are basically concerned with the behavior and forces between two particles. Using Democritis' static particle model, there was no way to understand how forces, locations, or directions could be communicated between the particles. Now the inward and outward spherical waves of the two particles provide that infomation continuously. The origin of the natural laws are described in a book, "*Exploring the Physics of the Unknown Universe*", by Dr. Milo Wolff. It is fascinating reading and the source of this technology of the wave structure of matter. ISBN 0-9627787-0-2

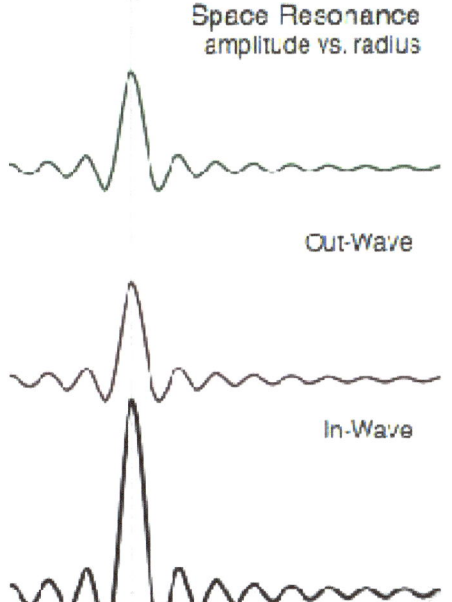

Superposition of In and Out waves creates Standing Wave

This figure graphs the amplitude of a Space Resonance along a radial line through the center.

The first two graphs show only the in-going and out-going portions of the resonance. Notice that the amplitudes are infinite at the center.

The bottom graph is the superposition of the inward and outward components, i.e. they are added together. This sum of the waves is the radial amplitude of the real electron. Notice that the amplitude at the center is finite.

The wave does not move inward or outward, it goes up and down. That is, it becomes a standing wave. Standing waves also occur for light and sound waves; for example on a violin string. The waves on the string can go up and down but not run along the string because of the posts. The electron, and other charged particles, are structured of standing spherical quantum waves.

The Particle vs. the Universe
Particles and the entire universe are interacting with each other through their inward and outward waves. Thus they become joined into one ensemble of waves which determines the behavior of the individual particles. The simplest example is Mach's Principle, which proposed (1890) that all the matter of the universe determines the law of inertia (F=ma).

What is an electron? Never in history has anyone been able to even imagine the structure of an electron. Until very recently, the electron was thought to be a 'point' made of an unknown, undefineable substance called 'charge'. Since a point has no dimensions, there is no structure and nothing to see.

The electron is a wave structure. Recently this structure was found to be two concentric spherical moving quantum waves. One wave moves inward and one wave moves outward. This spherical wave pair has the familiar properties of an electron which we call mass and charge, as well as all other properties. The illustration above shows the radial amplitude of these electron waves. These diagrams are mathematical plots of the IN and OUT waves and the SUM of both waves. The plots are the amplitudes of the waves drawn along a radius out from the electron center (at the left side). You will see that the farther from the center, the smaller the amplitudes become. This is like familiar sound or light waves or even water waves which become weaker as they move outward. Quantum waves are very tiny. In this diagram they are amplified a hundred, million, million times (10^{14}). You have to imagine them smaller - but that is not difficult.

Notice the behavior of the sum wave when the radius is near zero. The amplitude of the wave is not infinite at zero radius. Instead you notice it reaches a size of about + or - one inch on the screen. This is an important difference between the wave structure theory compared to the old point theory of the electron. In the point theory, electric amplitudes became infinite at zero radius. This is wrong because laboratory measurements show finite amplitudes. The wave structure, as shown here, gives the right result. Since the electrons and atoms are built of zero-point waves, they persist forever just like a magnetic force persists forever. It is postulated, eternal energy; it doesn't run down in a cascade fashion the way mechanical, electrical or chemical energy does.

Where are the orbits? For many (too many) years people imagined atoms as point electrons orbiting around a nucleus. This myth, obviously imitating our planetary system, was shown wrong by quantum theory more than sixty years ago. For example in the hydrogen atom, quantum theory predicts the electron presence as a symmetrical spherical cloud around the proton. Some physicists concluded that the point bits of matter were still there, even though quantum theory contains no notion of point particles. The old myth dies hard!

Actually, in the hydrogen atom both the electron wave-structure and the proton have the same center. The electron's structure can be imagined like an onion - spherical layers of waves around a center. The amplitude of the waves decreases like the blue standing wave in the bottom diagram. There are no point masses - no orbits, just waves.

The wave structure of matter illustrates a newly recognized essential property of a particle - communication. In order for particles to interact with each other, it is essential they communicate their position, direction, and presence to each other. The dynamic wave-structured electron can do this. The old static point particle cannot. It is wrong.

COSMIC MECHANICS

The great philosophers of cosmology, Clifford, Mach, Einstein, Wyle, Dirac, Schroedinger, have pointed out that only a wave structure of matter (particles) can conform to experimental data and fulfill the logic of reality and cosmology. This Quantum Wave Structure of Matter (WSM) has been found and is the origin of the natural laws. Since the WSM provides a quantitative origin of the fundamental natural laws, it is the basis of the physical sciences.

Only three basic Principles of Nature determine the wave medium and enable quantitative calculation of the WSM and the origin of the natural laws. The medium of the waves is space. The thought universe is a life-static, utterly devoid of matter, energy, space and time. It is totally still, but beings in the thought universe can disturb the stillness by making zero-point waves. The properties of particles (wave structures) and the laws embedded in them are derived from the properties of the life-static. This single entity, described by three principles, underlies everything. For the electron, the structure is a pair of spherical outward and inward quantum waves, convergent to a center, existing in ordinary Physical Universe space and termed a 'space resonance'. Every space resonance shares its binary wave structure pair with all others in our universe. Thus we exist in an inter-connected binary universe communicating at superluminal speeds. This knowledge is the basis of all miracles.

The predictive power of the WSM shows how the electron's wave pair is the physical origin of the previously unknown quantum spin. These two waves are a Dirac spinor satisfying the theoretical Dirac Equation. Spin occurs because the inward quantum wave continually undergoes spherical rotation at the center transforming to the outward wave.

The origins of the laws of the Physical Universe has been sought for centuries. Finding all of them at the same time and place is a philosopher's dream come true. It is of great importance to Science because the natural laws and quantum spin determine the structure of the Atomic Periodic Table, which dictates the varied forms of matter: metals, crystals, semi-conductors, and the molecules of life. The deep understanding of basic physics that is revealed opens a door to broad fields of applied technology such as integrated circuits, biophotonics, and commercial energy. It reveals a universe of real quantum wave structures that we live in but seldom see.

Finding these origins was very simple. The ancient Greek notion of a point particle was replaced with a *spherical wave structure*, which had already been predicted by Clifford, Einstein and Irwin Schroedinger, a century ago. After World War II enthusiasm for wave structure waned for 40 years. It was not complex or difficult but it was <u>different</u> because it disagreed with human emotional experience of point particles that most people were satisfied with. Emotions play a powerful, often hidden role in the motivation for research. We now have a new paradigm for understanding the Physical Universe and the underlying thought universe.

The golden balls to the left are not the 21 gram orbs which reduce our awareness. These golden balls are used by the thought universe entity that built the cat from its DNA code to define the symmetry of the skeleton and physical features of the organism. Human bodies have these golden balls also.

We don't easily see the space medium because our survival as an animal species depends mostly on our ability to fight with other animals seeking food, and to compete for mates that produce children, not closely related to the quantum field. In our self-focused human perspective few of us are even aware of the wave medium in which we exist. For survival, it doesn't matter what space is, or whether we can observe it - it exists nevertheless. This situation is much like the life of a fish who cannot comprehend the existence of water because it is too deeply immersed in it.

Like the fish, traditional science has tended to comprehend the universe in terms of our own local experiences. It was assumed that matter particles are like tiny grains of sand. As microscopy improved, models were created subdividing grains into other 'grains' - ad infinitum. But only a few people like Einstein saw that grains were impossible or ever asked how the grains communicate forces - unexplained by the old science. Formerly, forces were accepted as faith in Nature. Human perspective has many biases. We tend to see space as three rectangular dimensions, one of which is the vertical gravity vector of Earth, plus two other vectors perpendicular to it, shaped like the houses we live in. But in the cosmos, the shape of the enormous universe is *spherical* whose important dimensions are *inward* and *outward*, the direction of waves in space. In the vast expanse of the real universe, gravity occurs so rarely, that its direction is inconsequential in the larger scheme of things, despite its local importance to us. That is why the cat orbs (above) take the spherical form of creation. Cosmological space has spherical symmetry like suns, moons, asteroids and atoms.

The proof of the WSM is that all the natural laws can be obtained mathematically from the three basic principles describing the wave space medium. This wave structure of matter is simple and matches experimental measurements. In contrast, conventional physics required dozens of assumptions plus many more arbitrary constants to explain the operation of the laws. Even then some laws, like spin, were a puzzle with no origin known. On the atomic scale electromagnetism is inaccurate. The comparison of the old physics with its new wave structure can be compared to the theory of epicycles of the planets around the Earth before Copernicus and Galileo found that the planets traveled around the Sun.

In ancient times, men had little understanding of the world they lived in. They were mystified by the sun, which rose and set each day and gave them heat and light. Rain was water that helped crops to grow and soothed men's thirst. But sometimes rain also caused floods that destroyed towns. Because they understood so little of natural occurences ancient people created gods and goddesses to represent different parts of nature. For example, when ancient men heard the thunder of lightning, they decided that the sky was angry with them; so they imagined a god of thunder who was making the heavens shake with his anger. There were gods of the sea, the Earth, and the sun. There were even gods of ideas and gods of emotions – gods of love and gods of wisdom. The gods showed their anger by making storms or harsh winters. They showed their pleasure by providing good harvests and good hunting. These were the myths about gods and goddesses. And now we have the understanding of the wave structure of matter. Milo Wolff has finally closed the gap between the spiritual and the physical. He has shown us the energy exchanges between the creator and the created. And it turns out that we are the creators. Now we know! And we have the mathematics to apply it. We have an applied religious philosophy which will end the energy shortage, create new materials, improve health and take us to the stars just as our reptilian ancestors did.

82) THE NEW PARADIGM OF DR. MILO WOLFF

I now paraphrase the teachings of Dr. Milo Wolff according to my understanding. The discovery of the wave structure of the electron and other matter was a wonderful adventure to find the origin of the natural laws, a new powerful tool of technology, and an exciting window on science, cosmology, and ourselves. But some classic concepts must be discarded. For instance, a quantum physicist may expect that all quantum phenomena <u>must</u> derive from Schroedinger's Equation. Instead, it is the other way around; Schroedinger's Equation is derived from the quantum wave structure of the universe.

Some concepts must be changed such as the meaning of charge and mass, formerly assumed to be inherent properties of each independent particle. Instead of this assumption we find, as Schroedinger deduced, that location, charge and mass *are properties of the wave structure* and ultimately of the wave medium - space. This can now be proved.

Understanding our misconceptions comes from anthropology, which teaches that the quantum wave universe is not as helpful to survival of our personal genes as recognizing apples we can eat and avoiding tigers who want to eat us. We need to recognize those quickly. But it was not necessary that nature equip us to observe quantum waves quickly. Lacking personal experience of simple quantum waves, people chose to imagine that the electron is a "particle," like a bullet. Laboratory evidence does not support this human-oriented idea. Accordingly, belief must change from discrete particles to wave structure. But emotional rejection can occur if the new truth conflicts with established belief. Max Planck once said, "New scientific truth does not triumph by convincing its opponents, but because the opponents die and a new generation grows up unopposed to the new idea."

Figure 1. The Dynamic Waves of a Space Resonance

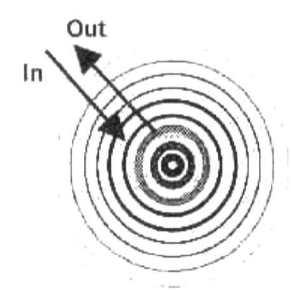

The resonance is composed of a spherical IN wave, which converges to the center, rotates to become an OUT wave, and diverges from the center. These two waves combine to form a standing wave whose peaks and nodes are like layers of an onion. The wave amplitude is a scalar number, <u>not</u> an electromagnetic vector. At the center, the wave amplitude is finite, not infinite, in agreement with observation.

Part II - Review of the Wave Structure of Matter

The wave-structured particle, Figure 1, is termed a *space resonance* (SR). The medium of the waves, and the leading player in the new scenario, is *space* that supposed void of which we formerly knew little. The properties of space resonances and the laws that they produce are derived from properties of space. Thus, this single entity, space, described by three principles, underlies *everything*.

Principle I - A Wave Equation

This Principle, an equation, describes how quantum waves are formed and travel in a space medium. The wave amplitudes are scalar numbers. If the medium is uniform, typical nearly everywhere, only *spherical waves* occur. If observed in relative motion, Doppler modulation and elliptical waves appear. If the medium is locally dense, as in the central region of a proton, waves *circulate* like sound waves in a drum or a sphere. Principle I is:

Quantum matter waves exist in space and are solutions of a scalar wave equation.

The wave equation is: $(grad)^2(AMP) - (1/c^2) d^2(AMP)/dt^2 = 0$

Where AMP is a scalar amplitude, c is the velocity of light, and t is the time. Its solutions, Figure 1, are a *pair* of spherical in/out waves that form the simple structure of the electron or positron — an oscillator. The waves decrease in intensity with increasing radius, like the forces of charge and gravity.

There are *two* combinations of the two *basic* in/out waves that form electrons and positrons, with opposite phase and spin rotation. Thus matter is constituted of two *binary* elements - like computer hardware. Although the variety of molecules and materials populating the universe is enormous, the basic building bricks are just two. Is there a profound meaning to this?

As is true for all oscillators in Nature the properties of the waves, such as speed and amplitude, derive from properties of their medium. Formation of the *space* medium is given by Principle II below.

Origin of the Natural Laws

The two combinations contain all experimental electron-positron properties. Charge depends on whether there is a + or - amplitude of the IN wave at the center. If a resonance is superimposed upon an anti-resonance, they annihilate. The amplitude at the center is finite as observed, not infinite as in the Coulomb rule. They obey Feynman's Rule: "A positron is an Electron going Backward in Time."

The properties of quantum mechanics (QM) and special relativity (SRT) are the result of the motion of one space resonance relative to another, which produces a *Doppler shift* in both the IN- and OUT- waves. All parameters of QM and SRT for a moving particle; that is, the de Broglie wavelength of QM and the relativistic mass and momentum changes, appear as algebraic factors in the Doppler-shifted waves exactly as experimentally measured.

Energy Transfer and the Action-at-a-Distance Paradox

A necessary new concept is the mechanism of energy exchange. Experience tells us that communication or acquisition of knowledge of any kind occurs only with an *energy transfer*. Storage of information, whether in a computer disk or in our mind, always requires an energy transfer. Energy is required to move a needle, to magnetize a tape, to stimulate a neuron. There are no exceptions. This rule of nature is embedded in biology and our instruments. Finding the energy transfer mechanism between particles is essential to understanding the natural laws. It is found naturally in Principle II of the WSM.

A major deficiency of the classical force laws is that they have no theoretical or physical *mechanism* for energy transfer. The formulas contain only constants, "mass" and "charge," - no mechanism. This was an inherent defect of the static point particle model. Einstein, Wheeler and Feynman knew this, recognizing that there must exist a continual dynamic means for forces to transfer energy and sought it in electromagnetic waves. Unfortunately there are no spherical solutions of the *vector* e-m wave equation. Hence the mechanism had to await the *scalar* waves of the WSM. We will see that wave communication is the means.

Ernst Mach's observation in 1883 was the first hint of the mechanism of cosmological energy transfer. He noticed that the inertia of a body depended on the presence of the visible stars. He asserted: *"Every local inertial frame is determined by the composite matter of the universe"* and jokingly, *"When the subway jerks, it is the fixed stars that throw you down."* His deduction arose from two different methods of measuring rotation. First, without looking at the sky one can measure the centrifugal force on a rotating mass m and use the inertia law F = ma to find circumferential speed and position, as in a gyroscope. The second method is to compare the object's angular position with the fixed (distant) stars. Both methods give exactly the same result!

Another way to look at this is to envision a Focault Pendulum. At the observatory the Focault pendulum has a fixed plane of oscillation. As the Earth turns beneath it, it knocks down dominoes through out the day. If we apply relativity concepts to this scene and hold the Earth still while we rotate the heavens around the pendulum – we get the same result.

Mach's Principle was criticized because it appeared to predict instantaneous *action-at-a-distance* across empty space. How can information travel from here to the stars and back again in an instant? Again the answer had to await the WSM where Nature's energy exchange mechanism, formerly unknown, is now seen as the interaction of waves in an ever-present universal medium. Space is not empty because it is a quantum wave medium created by oscillating waves from every particle in the universe (Principle II below). Inertia, charge, and other forces are mediated by the pervasive space medium. There is no need to travel across the universe.

Principle II - Space Density Principle (SDP)

This principle defines the quantum wave medium - space. It is fundamentally important because the properties of waves depend on properties of their medium. But, since the natural laws depend on the waves we deduce that the natural laws in turn depend on the medium. Thus, the cosmic being who's electrical field extends to infinity in space is the wellspring of everything.

Principle II is:

At each point in space, waves from all particles in the universe combine their intensities to form the wave medium of space.

The medium = space density ~ mc^2 = hf = $k'[\text{SUM OF}:\{(AMP_n)^2 \times (1/r_n^2)\}]$

In other words, at every point in space, the frequency f or the mass m of a particle depends on the sum of squares of all wave amplitudes AMP_n from the N particles inside the "Hubble universe", whose distance decreases inversely with their range r_n squared. The "Hubble Universe" is of radius $R = c/H$, where H is the Hubble constant.

This principle is a quantitative version of Mach's Principle because the space medium is the inertial frame of the law F = ma. When mass or charge is accelerated, energy exchange takes place between it and the surrounding space medium. In hindsight, this is the mechanism of charge radiation, unsuccessfully sought by Wheeler and Feynman (1945), attempting to use e-m waves instead of quantum waves.

Because there are a large number of particles, $N \sim 10^{80}$, in the Hubble universe, the medium is nearly constant everywhere and we observe a nearly constant speed of light. But near a large astronomical body like the Sun, the larger space density produces a measurable curvature of the paths of the inward and outward waves and thus of light and the motion of matter. We observe these as the effect of *gravity* described by Newton and also by the *space curvature* of Einstein's general relativity. Orbs also reflect incident light and refract it due to their increased space density compared to the air next to it.

The Energy Exchange Mechanism and Charge
Note that the self-waves of a resonance are counted too. Thus space becomes dense near the resonance centers due to their own large wave amplitude. Dense space at the central region propagates waves non-linearly, which allows energy transfer or *coupling* between two resonances. We observe this process and call it 'charge.' But there is no 'charge substance' involved. It is a property of the wave structure at the center.

Can this mechanism be tested? Yes. If a resonance's self-waves can be dominant in its space, then at some radius, r_o from the center, self-wave density must equal the total density of waves from the other N particles in the Universe. Evaluating this equality yields $r_o^2 = R2/3N$

The best astronomical measurements, $R = 10^{26}$ meters, $N = 10^{80}$ particles, yield $r_o = 6 \times 10^{-15}$ meters. This should be near the classical radius, $e2/mc2$ of an electron, which is 2.8×10^{-15} meters. The test is satisfied.

This is called the *Equation of the Cosmos;* a relation between the 'size' r_o of the electron and the size R of the Hubble Universe. Astonishingly, it describes how all the N particles of the Hubble Universe create the space medium and the 'charge' and 'mass' of each electron as a property of space.

Principle III - Minimum Amplitude Principle (MAP)

This third principle can be obtained from Principle II, but because it is a powerful law of the universe, which determines how interactions take place and how wave structures will move, I write it out separately:

***The total amplitude of all particle waves in space everywhere* always seeks a minimum.**

This principle is the disciplinarian of the universe. That is, energy transfers take place and wave-centers move in order to minimize total wave amplitude. Amplitudes are additive, so if two *opposite* resonances move together, the motion will minimize total amplitude. This explains empirical rules like, "Like charges repel and unlike charges attract" because those rules minimize total amplitude. The origins of many rules are now understood. For example, MAP produces the *Heisenberg Exclusion Principle*, which prevents two identical resonances (two fermions) from occupying the same state. This is not allowed because total amplitude would be a maximum, not a minimum. The operation of MAP is seen in ordinary situations like the water of a lake, which levels itself, and in the flow of heat that always moves from a hot source to a cold sink.

The Conservation of Energy

The transfer mechanism between combinations of resonances is a result of the dense (non-linear) space at resonance centers, which permits coupling or exchanges of wave frequency. When the waves of a potential source and a potential receiver, pass through both centers in an allowed transition, MAP minimizes the total of both amplitudes. In the source, the frequency (energy) of a wave state shifts downward. In the receiver, there is an equal shift upward. Only wave states (oscillators) with equal frequencies 'tuned' to each other can couple and shift frequency. Accordingly, the frequency (energy) changes must be equal and opposite. This is exactly the content of the *Conservation of Energy law*. The origin of this universal law is reduced to the matching of waves of two particles - not too different from tuning up an orchestra matched to the 'A' played by the first violin!

The Origin of the IN Waves and the Response of the Universe

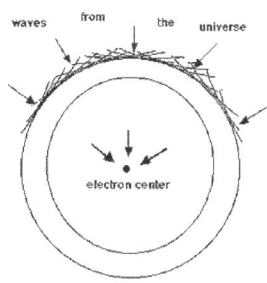

Figure 2. Formation of in-waves. The out-wave of every particle interacts with other matter in the universe. The response to the outgoing wave is Huygens wavelets from other matter that converge back to the center of

the initial out-wave. When the wavelets approach the center, their combined amplitude becomes larger, forming the IN wave. Thus every particle depends on all other particles for its existence. ***Our bodies are part of the universe and the universe is part of our bodies.***

At first thought, it is puzzling where the IN waves come from. This puzzle is our own fault - a result of looking at the waves of <u>only one particle</u>, and ignoring the waves of all other particles in space — over-simplification! To find reality, we must deal with the *real* wave-filled universe. When we study this question we find a rational origin of the inward waves:

Two hundred years ago Christian Huygens, a Dutch mathematician, found that if a surface containing many separate wave sources was examined at a distance, the combined wavelets appeared as a single wave front having the shape of the surface. This wave front is termed a 'Huygens Combination' of the separate wavelets (Figure 2). This mechanism is the origin the in-waves, as follows: When an outgoing wave encounters other particles, their out-waves are joined with part of the initial out-wave. Then out-waves from all other particles can form a Huygens Combination wave front that is the in-wave of the initial particle. These waves arrive in phase at the initial center. This occurs throughout the universe so that every particle depends on all others to create its in-wave.

We see it is wrong to imagine each particle as one pair of in- and out-waves, because one pair cannot exist alone. We have to think of each particle as inextricably joined with other matter of the universe. Although particle centers are widely separated, all particles together are one unified structure. Thus, ***our bodies are part of a unified universe and the universe is part of our bodies.***

Part III - The origin of the electron's spin

As an example of the depth of understanding and universality of the Wave Structure of Matter, Milo described for the first time the origin of the spin of the electron. The physical nature and cause of electron spin has been sought for 75 years ever since a *theoretical* theory of spin had been developed by Nobel laureate Paul Dirac in 1926. He theoretically predicted the *positron*, found in cosmic rays five years later by C. D. Anderson.

The Physical Mechanism of Spin

Spin occurs when the in-wave arrives at the center and *rotates* continuously in order to transform into the out-wave. To accomplish rotation there are strict (boundary) conditions on the amplitudes and polarity of the IN and OUT waves. Rotation cannot be allowed to twist up space without limit. The spherical wave amplitudes must continually and smoothly change while changing direction of motion. The in-wave amplitude at the center must be equal and opposite to the out-wave (one half cycle different).

It turns out this happens using a known property of 3D space called *spherical rotation* in which space rotates continually around a *point* and returns to its initial state after two turns. In spherical rotation there is no fixed axis like *cylindrical* rotation of a wheel. Spherical symmetry is preserved because the center of rotation is a *point*. One direction of rotation produces the electron, the other the positron. This is why every charged particle has an anti-particle.

Batty-Pratt & Racey (1980) analyzed spherical rotation and showed that an exponential oscillator, eiwt, was a spinor. Milo Wolff realized in 1989 that the exponential in-out oscillator waves of the WSM were the *real* physical spinors satisfying the Dirac Equation. It is significant that only 3D space has this remarkable property. If this real property of 3D space did not exist, particles and matter could not exist. Life and the universe as we know it could not exist.

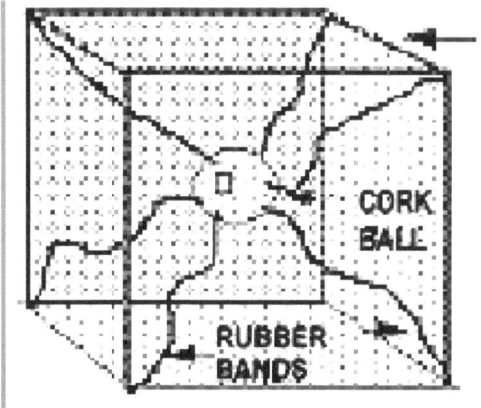

SUPPORTING FRAME OF STICKS

Figure 3. A Model of Spherical Rotation

This apparatus is easily made out of a few sticks, a cork, and six rubber bands. The cork can be rotated (taking care not to knot up the rubber bands) continuously without entangling the rubber bands! The cork and bands will return to their initial configuration every two turns. This demonstrates a little known variety of rotation. It has application to particle theory because spherical rotation does not destroy continuity of space.

A Model of Spherical Rotation

Spherical rotation in 3D space can be modeled by a ball held by threads inside a cubical frame shown in Figure 3. The threads represent the coordinates of the space and the rotating ball represents the space at the center of the converging and diverging quantum waves. The ball can be turned about any axis starting from any initial position. If the ball is rotated continuously it returns to its initial configuration after every two rotations.

Connecting Quantum Theory and Relativity

Before the WSM, there had been no known physical reason for the theoretical mass increase of relativity. Likewise there were no physical reasons for quantum theory or spin. Were these apparently separate laws connected or not? Indeed, many theorists proclaimed that these phenomena were irreconcilable! Few had thought about a connection because most physicists were satisfied with the separate and reasonably accurate theoretical reasons for each.

Dirac's work was a clue that they are connected because spin, relativity, and QM were joined in Dirac's work, albeit theoretically. The WSM now reveals their simple physical connection - Doppler effects.

The increase of mass (energy or frequency) $m = m_o[1 - v^2/c^2]^{-1/2}$ of a space resonance is due to the Doppler increase of frequency seen by a moving observer. This immediately results in the conservation of energy equation used by Dirac:

$$E^2 = p^2c^2 + m_o^2c^4 \quad (1)$$

This energy equation is also found using the wave perspective because E and p are superimposed waves. In engineering the total intensity of two waves is given by the sum of their squares. Likewise, the de Broglie wavelength $L = h/p$ is also a Doppler change of wavelength seen by a moving observer and it leads to the Schrodinger Equation. Thus the union of spin, mass increase and quantum theory occurs in the spherical rotation at the wave center, and is part of the WSM in general.

Part IV - Conclusions

We can have confidence that the Wave Structure of Matter is the true physical reality of the universe. The required proof of the WSM is that the experimental evidence, which empirically founded the natural laws, must necessarily agree with the laws predicted by the WSM. It does.

The philosophical conclusions are fascinating, particularly concerning the connectedness of the universe: *Everything we are and observe here on Earth, matter-laws-life, necessarily depends on the existence of all the matter in the universe.* We must conclude, if the stars and galaxies were not in the heavens, we could not exist! *Thus, we are part of the universe, and the universe is part of us*. (ie., our bodies, not our spirits)

But the practical value of the WSM theory is the insight it provides. It allows scientists to deeply analyze quantum wave structures, the cosmos, and the natural laws. In the R&D laboratory, the new insight should advance electronic applications, especially IC and memory devices because their tiny transistor elements use quantum effects to control the flow of currents.

The new knowledge should improve communication and the efficiency of energy transmission. For example, conduction of electric energy along a wire is a quantum energy transfer process between SRs. Knowing this, energy losses may be reduced, lowering cost and increasing transmission distances.

What is the mechanism of these matter wave connections? Each atom of our body structure

arranges its waves to form minimum amplitudes that satisfy MAP. In the process, each of its neighbors makes a contribution to its in-wave structure. If one atom is disturbed, others must be affected. This is an elementary information transfer. Since every life cell (each particle) contains the holographic image of waves from other parts (other particles) of the body, each cell is a memory device. So the apparatus of computers appears to be there in the body's mind. The computer has a biophotonic wiring system, the clock speed is the frequency of the electron, there are 52 sensory inputs, thousands of outputs to the motor controls, an orb for a central processing unit, muscles and organs as peripheral output devices, and a metabolic power supply.

So there you have it. Now you know how we create reality with zsro-point and quantum waves. Now you can think outside the box.

83) APPLICATIONS OF SPACE RESONANCES

Now that Dr. Wolff has discovered the physics of the wave structure of matter, what can we do with it? Here are some applications.

Relativity Drive From New Scientist Magazine:

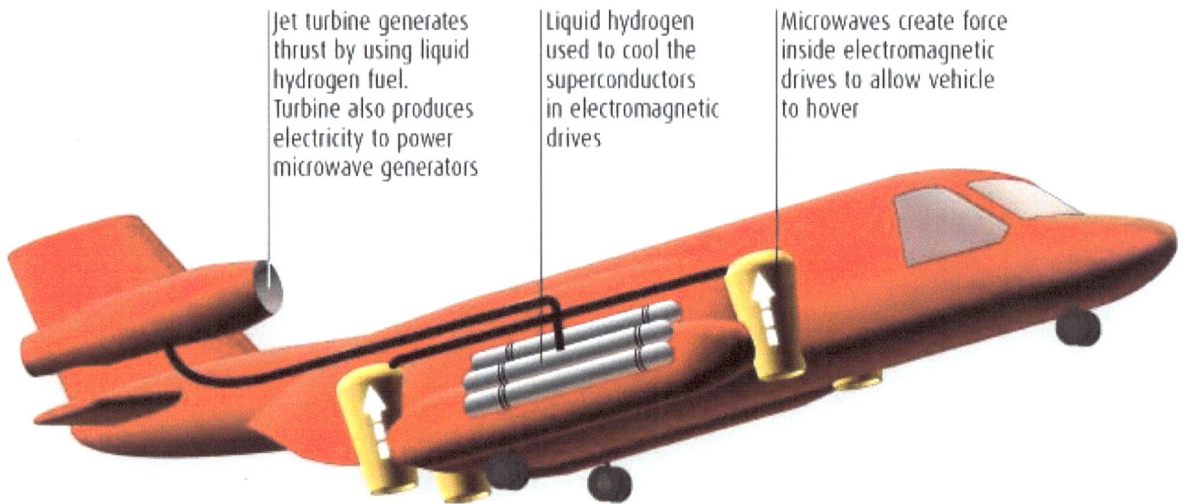

LOOK, NO WINGS!
If Roger Shawyer's electromagnetic drive performs as he hopes, it might be possible to build vehicles that hover. Liquid hydrogen could cool the drives and fuel a turbine to propel the craft forwards

Jet turbine generates thrust by using liquid hydrogen fuel. Turbine also produces electricity to power microwave generators

Liquid hydrogen used to cool the superconductors in electromagnetic drives

Microwaves create force inside electromagnetic drives to allow vehicle to hover

Roger Shawyer has developed an engine with no moving parts that he believes can replace rockets and make trains, planes and automobiles obsolete. "The end of wings and wheels" is how he puts it. It's a bold claim..

Of course, any crackpot can rough out plans for a warp drive. What they never show you is evidence that it works. Shawyer is different. He has built a working prototype to test his ideas, and as a respected spacecraft engineer he has persuaded the British government to fund his work. Now organisations from other parts of the world, including the US Air Force and the Chinese government, are beating a path to his tiny company.

The device that has sparked their interest is an engine that generates thrust purely from electromagnetic radiation - microwaves to be precise - by exploiting the strange properties of relativity. It has no moving parts, and releases no exhaust or noxious emissions. Potentially, it could pack the punch of a rocket in a box the size of a suitcase. It could one day replace the engines on almost any spacecraft. More advanced versions might allow cars to lift from the ground and hover. It could even lead to aircraft that will not need wings at all. I can't help thinking that it sounds too good to be true.

Shawyer had a bright idea. He wanted to build a replacement system for the small thrusters conventional satellites use to stay in orbit. The fuel they need makes up about half their launch weight, and also limits a satellite's life: once it runs out, the vehicle drifts out of position and must be replaced. Shawyer's engine, by contrast, would be propelled by microwaves generated from solar energy. The photovoltaic cells would eliminate the fuel, and with the launch weight halved, satellite manufacturers could send up two craft for the price of one, so you would only need half as many launches.

Electromagnetic waves like sunlight and microwaves striking a suface produce incredibly small thrust forces. What if you could amplify the effect? That's exactly the idea that Shawyer stumbled on in the 1970s while working for a British military technology company called Sperry Gyroscope. Shawyer's expertise is in microwaves, and when he was asked to come up with a gyroscopic device for a guidance system he instead came up with the idea for an electromagnetic engine. He even unearthed a 1950s paper by Alex Cullen, an electrical engineer at University College London, describing how electromagnetic energy might create a force. "It came to nothing at the time, but the idea stuck in my head," he says.

In his workshop, Shawyer explains how this led him to a way of producing thrust. For years he has explored ways to confine microwaves inside waveguides, hollow tubes that trap radiation and direct it along their length. Take a standard copper waveguide and close off both ends. Now create microwaves using a magnetron, a device found in every microwave oven. If you inject these microwaves into the cavity, the microwaves will bounce from one end of the cavity to the other. According to the principles outlined by Maxwell, this will produce a tiny force on the end walls. Now carefully match the size of the cavity to the wavelength of the microwaves and you create a chamber in which the microwaves resonate, allowing it to store large amounts of energy.

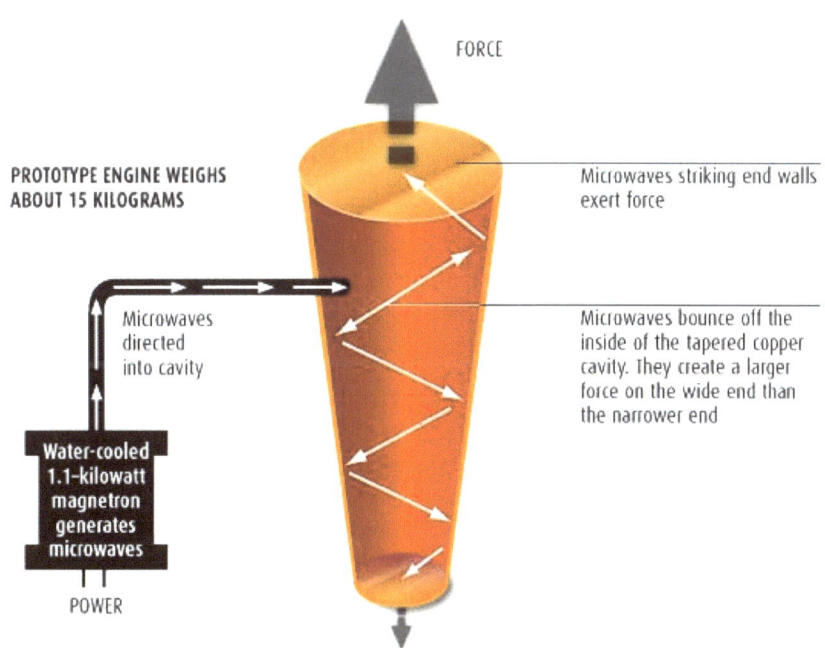

THE ELECTROMAGNETIC DRIVE

Microwaves trapped in a cavity exert a force on the end walls. By making the area of one end greater than the other, Roger Shawyer says he can tailor this force so his device generates thrust

How the electromagnetic drive compares

Engine	European Space Agency's SMART-1 ion engine	Electromagnetic drive
Power required	700 watts	700 watts
Thrust generated	70 millinewtons	88 millinewtons
Operational life	1.6 years	15 years
Weight	94 kilograms	9 kilograms

What's crucial here is the Q-value of the cavity - a measure of how well a vibrating system prevents its energy dissipating into heat, or how slowly the oscillations are damped down. For example, a pendulum swinging in air would have a high Q, while a pendulum immersed in oil would have a low one. If microwaves leak out of the cavity, the Q will be low. A cavity with a high Q-value can store large amounts of microwave energy with few losses, and this means the radiation will exert relatively large forces on the ends of the cavity. You might think the forces on the end walls will cancel each other out, but Shawyer worked out that with a suitably shaped resonant cavity, wider at one end than the other, the radiation pressure exerted by the microwaves at the wide end would be higher than at the narrow one.

Key is the fact that the diameter of a tubular cavity alters the path - and hence the effective velocity - of the microwaves travelling through it. Microwaves moving along a relatively wide tube follow a more or less uninterrupted path from end to end, while microwaves in a narrow tube move along it by reflecting off the walls. The narrower the tube gets, the more the microwaves get reflected and the slower their effective velocity along the tube becomes. Shawyer calculates the microwaves striking the end wall at the narrow end of his cavity will transfer less momentum to the cavity than those striking the wider end (see Diagram). The result is a net force that pushes the cavity in one direction. And that's it, Shawyer says.

Hang on a minute, though. If the cavity is to move, it must be pushed by something. A rocket engine, for example, is propelled by hot exhaust gases pushing on the rear of the rocket. How can photons confined inside a cavity make the cavity move? This is where relativity and the strange nature of light come in. Since the microwave photons in the waveguide are travelling close to the speed of light, any attempt to resolve the forces they generate must take account of Einstein's special theory of relativity. This says that the microwaves move in their own frame of reference. In other words they move independently of the cavity - as if they are outside it. As a result, the microwaves themselves exert a push on the cavity.

Each photon that a magnetron fires into the cavity creates an equal and opposite reaction - like the recoil force on a gun as it fires a bullet. With Shawyer's design, however, this force is minuscule compared with the forces generated in the resonant cavity, because the photons reflect back and forth up to 50,000 times. With each reflection, a reaction occurs between the cavity and the photon, each operating in its own frame of reference. This generates a tiny force, which for a powerful microwave beam confined in the cavity adds up to produce a perceptible thrust on the cavity itself.

Shawyer's calculations have not convinced everyone. Depending on who you talk to Shawyer is either a genius or a purveyor of snake oil. David Jefferies, a microwave engineer at the University of Surrey in the UK, is adamant that there is an error in Shawyer's thinking. "It's a load of bloody rubbish," he says. At the other end of the scale is Stepan Lucyszyn, a microwave engineer at Imperial College London. "I think it's outstanding science," he says. Marc Millis, the engineer behind NASA's programme to assess revolutionary propulsion technology accepts that the net forces inside the cavity will be unequal, but as for the thrust it generates, he wants to see the hard evidence before making a judgement.

Thrust from a box
Shawyer's electromagnetic drive - emdrive for short - consists in essence of a microwave generator attached to what looks like a large copper cake tin. It needs a power supply for the magnetron, but there are no moving parts and no fuel - just a cord to plug it into the mains. Various pipes add complexity, but they are just there to keep the chamber cool. And the device seems to work: by mounting it on a sensitive balance, he has shown that it generates about 16 millinewtons of thrust, using 1 kilowatt of electrical power. Shawyer calculated that his first prototype had a Q of 5900. With his second thruster, he managed to raise the Q to 50,000 allowing it to generate a force of about 300 millinewtons - 100 times what Cosmos 1 could achieve. It's not enough for Earth-based use, but it's revolutionary for spacecraft.
One of the conditions of Shawyer's £250,000 funding from the UK's Department of Trade and Industry is that his research be independently reviewed, and he has been meticulous in

cataloguing his work and in measuring the forces involved. "It's not easy because the forces are tiny compared to the weight of the equipment," he says.

Optimising the cavity is crucial, and it's as much art as science. Energy leaks out in all kinds of ways: microwaves heat the cavity, for example, changing its electrical characteristics so that it no longer resonates. At very high powers, microwaves can rip electrons out of the metal, causing sparks and a dramatic loss of power. "It can be a very fine balancing act," says Shawyer.

To review the project, the UK government hired John Spiller, an independent space engineer. He was impressed. He says the thruster's design is practical and could be adapted fairly easily to operate in space. He points out, though, that the drive needs to be developed further and tested by an independent group with its own equipment. "It certainly needs to be flown experimentally," he says.

To space and beyond
His plan is to license the technology to a major player in the space industry who can adapt the design and send up a test satellite to prove that it works. If all goes to plan, Shawyer believes he could see the engine tested in space within two years. He estimates that his thruster could save the space industry $15 billion over the next 10 years. Spiller is more cautious. While the engine could certainly reduce the launch weight of a satellite, he doubts it will significantly increase its lifetime since other parts will still wear out. The space industry might not need to worry after all.

BACK-ENGINEERING ALIEN TECHNOLOGY

"Last month (April 2007), my wife and I were on a walk when we noticed a very large, very strange "craft" in the sky. My wife took a picture with her cell phone camera (left). A few days later a friend (and neighbor) lent me his camera and came with me to take photos of this "craft". We found it and took a number of very clear photos. Picture #4 is taken from right below this thing and I must give my friend credit as I was not brave enough to get close enough to take this picture myself!

"The craft is almost completely silent and moves very smoothly. It usually moves slowly until it decides to take off. Then it moves VERY quickly and is out of sight in the blink of an eye. MORE THAN ANYTHING I simply want to understand what this is and why it is here?"

Dr. F. Lee Aeilts
--Chad

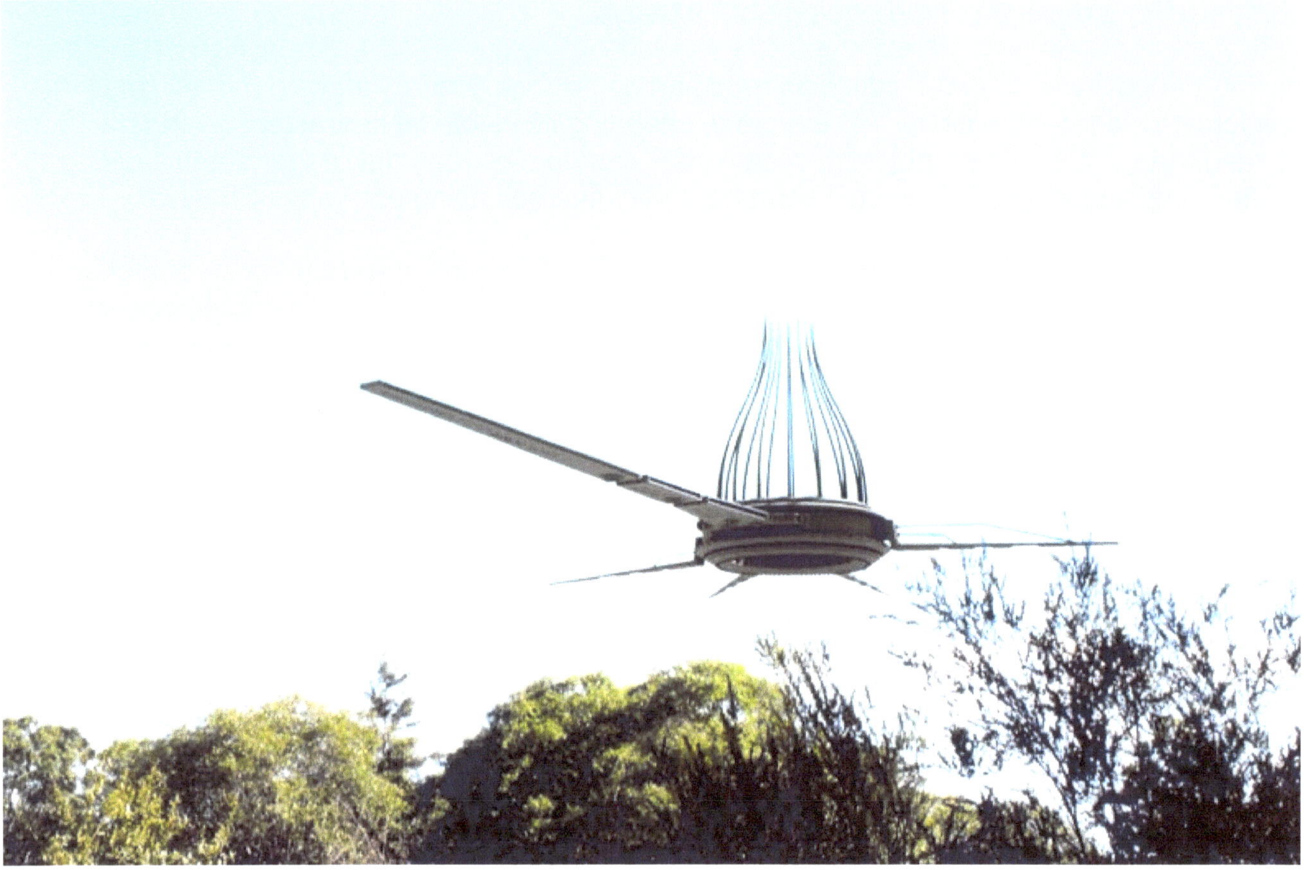

C ommercial
A pplications
R esearch for
E xtraterrestrial
T echnology

Q4-86 RESEARCH REPORT

COSMIC MECHANICS

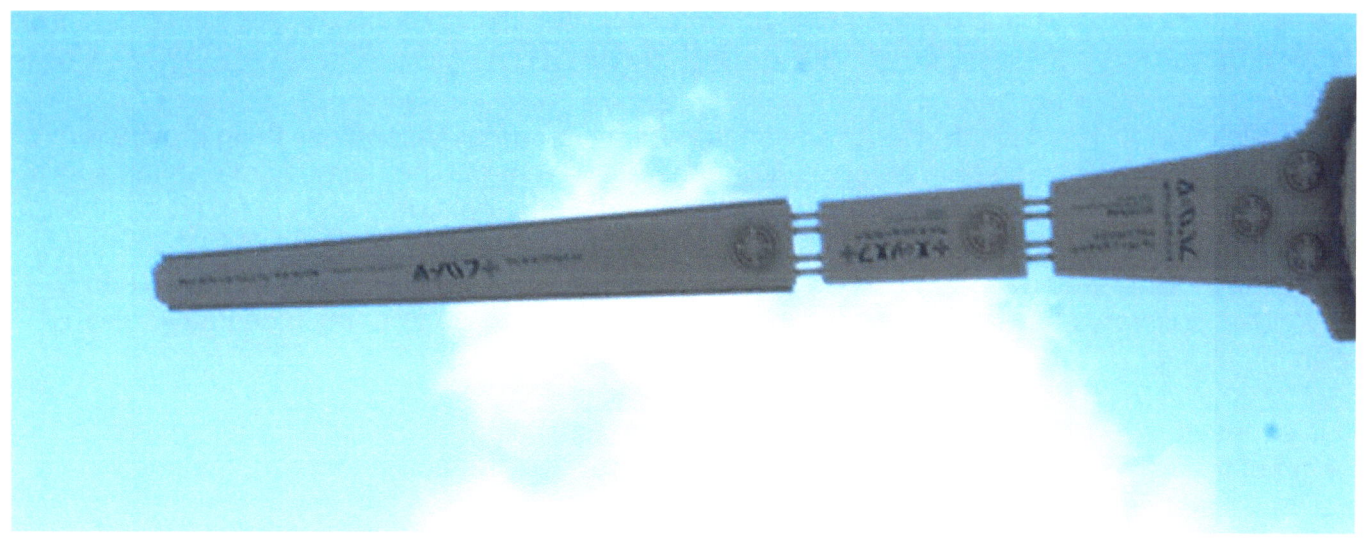

271

"I see this thing VERY often. Since it first appeared, I have probably seen this thing maybe 8 different times since the first appearance. My friend and I went out the next day after I first saw it to get the photos, but it was not there. Then we tried again the next day, and we found it within like 30 minutes and followed it for a while. Most of the time I see it out of windows in my house in the distance. But I would say almost half of the hikes I have gone on in my area, I have seen it very close. It is very easy to photograph and many neighbors aside from my friend have also seen it.

"It is almost totally silent but not quite. It makes kind of "crackling" noises. It's hard to describe them but they are only intermittent and not very loud, but you can notice them. Sometimes there is a very slight hum that sounds kind of mechanical, almost like when you are near very large power lines. But it is nothing loud like a jet engine. It is very quiet for the most part.

"It moves almost like an insect. If you have ever seen a bug on a pond, it is kind of like that. It is VERY smooth and slow most of the time, but then every now and then it will rotate very quickly and go VERY fast into another direction, then stop, and repeat the process all over again. There is just something very unnatural about the way it moves.

"Also, I have had maybe 4 headaches in the last week, and I am normally not the kind of person who really ever gets them. Also my wife has been tired and fatigued lately. She is about a month

pregnant, and the doctor said fatigue is normal around this time, but I worry that it is a lot."
--Chad

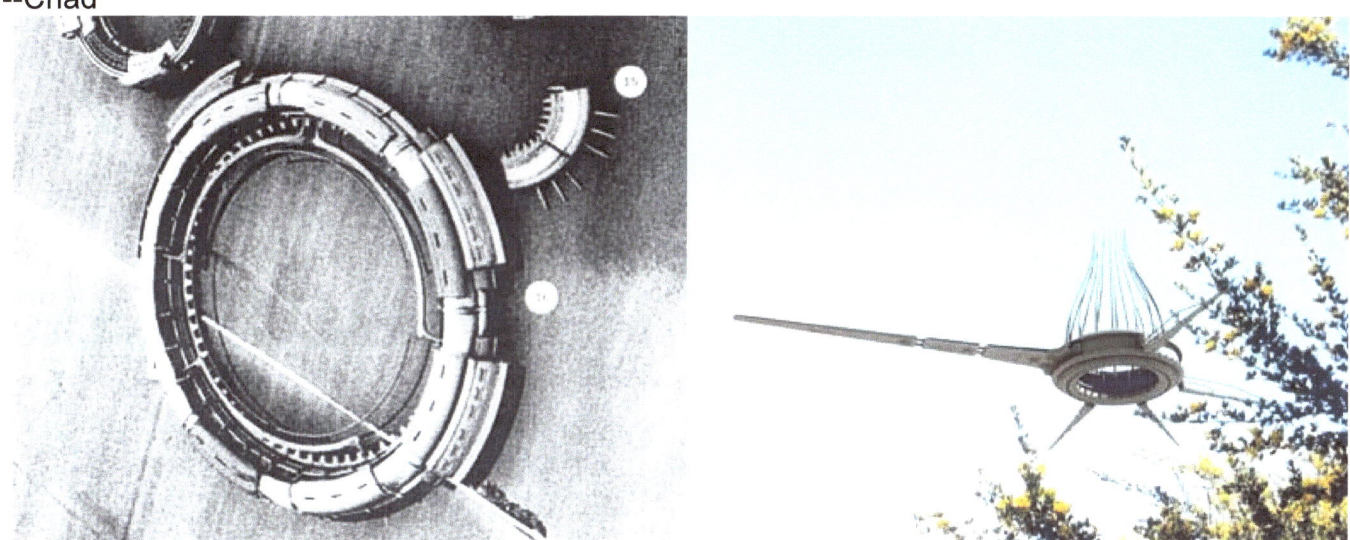

Isaac's 'CARET' image compared with Chad's drone photo

I have information that can explain a great deal of what is going on in the photos provided by Chad, Rajman, the Lake Tahoe witness, as well as the photos taken by Jenna and Ty in Big Basin, CA…

I'm going by the alias "Isaac" while I release this information. I'm an electrical engineer and computer scientist and used to work in a program called "CARET" that was concerned with research and development based on extra-terrestrial technology. Many key elements of the technology I worked with during my time with CARET are clearly visible in these crafts. This work was done in the 80's in Palo Alto, CA, so it's no surprise to me that these sightings are taking place within such proximity to that area.

Attached to this email you will find numerous scans of photographs and documents from my CARET days that pertain to what I'm going to tell you, and would like to share with your audience. In addition, I have put up a small website that has very high resolution copies of all of these scans, as well as a lengthy letter that explains a lot about me, the CARET program, and these sightings. I've been listening to your (Coast to Coast AM) show since 2002 and would like to work with you to get this information out there. Please read the text at the website for the rest of the details, as I am taking a fairly significant personal risk and would like to make sure it's worth it:
isaaccaret.fortunecity.com

All I ask in sharing this information is that it's kept together. There are 4 photos and a total of 15 pages of scanned documents. If you'd like to resize them for your website that is fine… I want people to have easy access to this information IN FULL. I feel this is reasonable and it is all I ask. I am NOT trying to sell anything and the only reason I created this very simple website is to ensure all the information is presented correctly and in one place. --Isaac

Dr. F. Lee Aeilts

PALO ALTO CARET LABORATORY   Q4-86 RESEARCH REPORT

1. OVERVIEW

This document is intended as a primer on the tentative findings of the Q4 1986 research phase (referred to herein as "Q4-86") at the Palo Alto CARET Laboratory (PACL). In accordance with the CARET program mission statement, the goal of this research has been achieving a greater understanding of extraterrestrial technology within the context of commercial applications and civilian use. Examples of such applications, in no particular order, include transportation, medicine, construction, energy, computing and communication. The ultimate goal of this research is to provide a core set of advanced technologies in a condition suitable for patent review.

2. EXTRACTION

The process of converting raw artifacts of extraterrestrial origin to usable, fully-documented human technology is termed *extraction*. The extraction process ultimately consists of two phases: first is the establishment of a complete theoretical and operational understanding of the artifact, and second is a distillation of the artifact's underlying principles into a usable, product-oriented technology. Suggestions of specific product applications on behalf of PACL have been encouraged, but are not considered mandatory or essential.

The results of a successful extraction are collected in what is termed an *extraction package* (EP), which should include the following:

1. Complete theoretical and operational overview
2. Assessment and summary of compositional materials
3. At least three (3) working prototypes, demonstrating multiple instances of successful, repeatable and reliable implementation
4. Assembly notes and BOM

At the time of this writing, a fully successful extraction has not yet been achieved, although numerous threads of research are showing promise.

Comprehensive documentation of PACL's extraction process can be found in document PACL-D0006, entitled "PACL Extraction Procedure Guide".

3. EXECUTIVE SUMMARY OF Q4-86

Q4-86 focused on four key subjects, all of which were based on artifacts of extraterrestrial origin obtained from crash site recovery operations conducted during the last two decades within the continental United States. These subjects are:

1. "Personal" antigravity generator (so-named for its small, portable size)
2. Three-dimensional image recorder/projector

3. A complex system of symbols and geometric constructs capable of both defining the functionality of certain artifacts as well as manipulating their behavior, crudely analogous to a computer programming language but without the need for a compilation or interpretation phase.

4. ███████████████████████████████████████

4. RESEARCH SUBJECT: "PERSONAL" ANTIGRAVITY

Antigravity technologies are among the most ubiquitous recovered from extraterrestrial crafts. While antigravity is most commonly associated with propulsion, the principles underlying the technology extend into a far broader domain; indeed, virtually all aspects of most extraterrestrial craft seem to incorporate its use in some way. A prominent example is the seemingly impenetrable field, of controllable diameter and attenuation, surrounding the craft that protects it from weather conditions and the surrounding environment, as well as debris, and, unsurprisingly, ballistic weaponry. Additional examples include dampening of G-force on passengers and on-board equipment, movement of doors and hatches (or their closest equivalents), and even placement of fixtures (such as control consoles, or their closest equivalents) within a given space. Perhaps most startling is the fact that the very components within a given extraterrestrial craft appear to be held in place, in relation to one another, exclusively by antigravitational means. This is a partial explanation for the commonly noted lack of rivets and adhesives in the construction of these crafts.

PACL aims to translate this technology into a product-oriented EP capable of direct application within the consumer market. However, since the sudden emergence of such radically advanced technology would undoubtedly yield destructive consequences, PACL recommends a strategy of incremental dissemination in which deliberately downgraded versions of the original technology are released over a period of years or decades to soften the impact of integration with existing infrastructures, in technological, economic and social terms.

4.1. WHAT IS PERSONAL ANTIGRAVITY?

Not all recovered extraterrestrial technologies are equal, and many previous experiments on antigravity have been performed on cumbersome artifacts suffering from enormous form factors and impractical weights. An ironic consequence of these previous generations of experimentation is that many man-made aircraft that would be otherwise ideal for antigravity propulsion models are incapable of supporting the weight of the device before its gravity-canceling effects are activated. This has lead to many clumsy and accident-prone solutions, such as using a second antigravity generator to load and position the first within the aircraft before activation and takeoff, and then repeating the process in reverse after landing but before deactivation. Despite some minor successes in narrowly-defined domains, these approaches are obviously not acceptable in the long term.

Recently, however, a rather different implementation of antigravity technology has appeared, undoubtedly the product of a different, and presumably more advanced source ██ it can produce gravity-canceling effects of magnitudes comparable to existing artifacts in a package less than two feet across and weighing less than five pounds.

PACL has termed this technology "personal antigravity", as its virtually negligible weight and dimensions suggest applications as focused as antigravity generation for a *single* human user. Early experiments suggest, however, that despite its remarkable precision and focus, this technology is equally effective when broadened to deal with massive payloads of arbitrary scales.

4.2. OVERVIEW OF RECOVERED ANTIGRAVITY ARTIFACTS

4.2.1. KEY ARTIFACTS

PACL has conducted the brunt of its antigravity research on three key artifacts. The first is what PACL considers to be an "antigravity generator" (seen in figure 4.1), a device that appears to provide a "source" of antigravity that can then be projected onto or harnessed by other components within the craft. The second two artifacts are curved I-beam segments (seen in figure 4.2) that, when placed anywhere within a certain radius of the generator during a specific mode of its operation, immediately fly into what is presumed to be their relative positions within the original construction of the craft.

The generator artifact is assigned the identification code *A1*. The I-beam artifacts are assigned identification codes *A2* and *A3*.

4.2.2. SECONDARY ARTIFACTS

Additionally, PACL has been provided with a small, ██████████████████ device capable of controlling A1 by activating and deactivating it, as well as switching between its three primary modes of operation. This device, assigned the identification code *S1*, is of particularly sensitive importance, as it is the only known method of controlling A1. ██

4.2.3. RIGID SPATIAL RELATIONSHIPS

Unlike the more general-purpose antigravity fields generated by implementations of this technology obtained from other sources, A1 is capable of multiple modes of operation and varying levels of precision. Perhaps the most compelling aspect of A1's functionality is its ability to focus its antigravitational effects on specific objects, rather than entire spatial volumes, creating what PACL has termed a *rigid spatial relationship* (RSR).

An RSR can be thought of as creating an "implicit solid" between two or more constituent parts separated by empty space. Once in effect, these constituent parts behave as if they

COSMIC MECHANICS

Figure 4.1
The artifacts used by PACL during the antigravity research phase of Q4-86.

are directly and physically linked, and are completely inseparable by pulling or pushing them in opposing directions. Only when the effect of A1 is deactivated will they once again behave as separate objects.

As an example, imagine cutting a broomstick into two segments, each one foot in length. Once separated, each segment is its own object, capable of being moved or rotated independently of the other. Under the effect of an RSR, however, the segments might behave as if they were a three-foot rod consisting of both foot-long broomstick segments separated by an additional foot of empty space. While the two rod segments would still appear to be separate, to the point that an observer would be able to pass their hand through the space that separates them, they would be unable to move one of the rods without the other behaving as if it were directly attached.

4.2.4. OVERVIEW OF A1

A1 consists of a two-segment cylindrical core, 1 foot, 2.2 inches in length and 8.3 inches in diameter, with needle-like appendages extending from each end. The total length of the device, with needles included, is 2 feet, 2.4 inches. Both core segments feature a triangular array of three "arms", extending 7.6 inches from the center of the core,

COSMIC MECHANICS

Figure 4.2
Close-up shot of the I-beam segments

each of which end in a circular "pad" with a diameter of 2 inches. The device weighs approximately 4 pounds, 3 ounces.

Research on the internal functionality of A1 began late in Q4-86, and as such, little is currently known. What is certain, however, is that the device contains no moving parts whatsoever, does not feature any kind of control interface in the form of buttons, switches, or levers, and, apparently, can only be manipulated by the technology contained in S1. According to the limited data to which PACL has been given access in regards to the placement and housing of A1 within the original craft, A1 was one of a pair of identical generators, together responsible for all antigravity-related functionality, from propulsion of the craft itself to placement of all components within the craft's internal design. From this information, as well as experiments conducted with S1, it has been discovered that A1 operates in one of at least three modes of operation:

1. *Field mode.* A1 generates a field of (presumably) arbitrary size and any shape that can be expressed as a convex volume. Within this field, gravity is effectively redefined with any desired strength and orientation. The parameters of this mode, including the shape of the field itself, are defined by ████████████████ ████████████ S1. Surprisingly, A1 does not appear capable of generating a field with any degree of concavity, nor can the strength or orientation of the artificial gravity within the field vary from one point to another. An example of

field mode would be creating a controlled gravity environment within an aircraft or spacecraft for passengers and cargo.

2. *Component mode.* Rather than generating a general-purpose field of constant gravity control, A1 will manipulate the gravitational effect on specific objects, allowing them to take any position or orientation relative to its own centroid. Component mode appears to be used commonly for maintaining the physical construction of a craft's design. Rather than attaching a craft's components to one another by way of rivets, adhesives, welding or the like, they are simply held in place, quite precisely, by antigravitational means. Unlike field mode, PACL has not yet been successful in controlling the parameters or data that drive this mode. S1 does not appear capable of controlling this mode beyond activating or deactivating it. Once in effect, the details of which components are affected, and how, seem to be provided by the components themselves. See the following section for more information. Component mode is responsible for the RSR effect described in the previous section and depicted in figure 4.4.

3. *Multi mode.* A1 combines the functionality of the field and component modes, producing specific antigravity effects on individual components while also generating any number of general-purpose gravity control fields. The same limitations that apply to the field generated in field mode apply to fields generated in this mode as well, but the ability to create multiple fields of differing parameters allows those limitations to be effectively circumvented in most situations. It is believed that this mode was used most commonly for managing the antigravitational needs of the original craft.

4.2.5. OVERVIEW OF A2 AND A3

On their own, A2 and A3 appear to be completely non-functional segments of a curved I-beam (seen in in figure 4.3). However, when A1 is switched into component mode, their position and orientation in relation to A1's centroid are precisely enforced with an RSR (seen in figure 4.4).

A2 and A3 are primarily differentiated by their lengths, which are 7.2 inches and 9.1 inches, respectively. Despite the difference in their lengths, both artifacts weigh approximately 2.6 ounces.

While initial experimentation indicated that the artifacts were composed of a consistent, solid material, experiments on A1's component mode suggest that the artifacts are more internally complex, somehow containing information that describes their position and orientation in relation to A1 when the mode is in effect. Whether or not they possess additional functionality beyond the storage of this information is currently unknown, but is considered likely due to their otherwise ambiguous purpose within the craft's design.

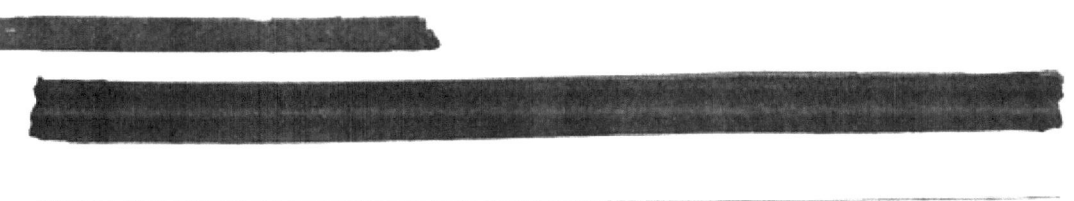

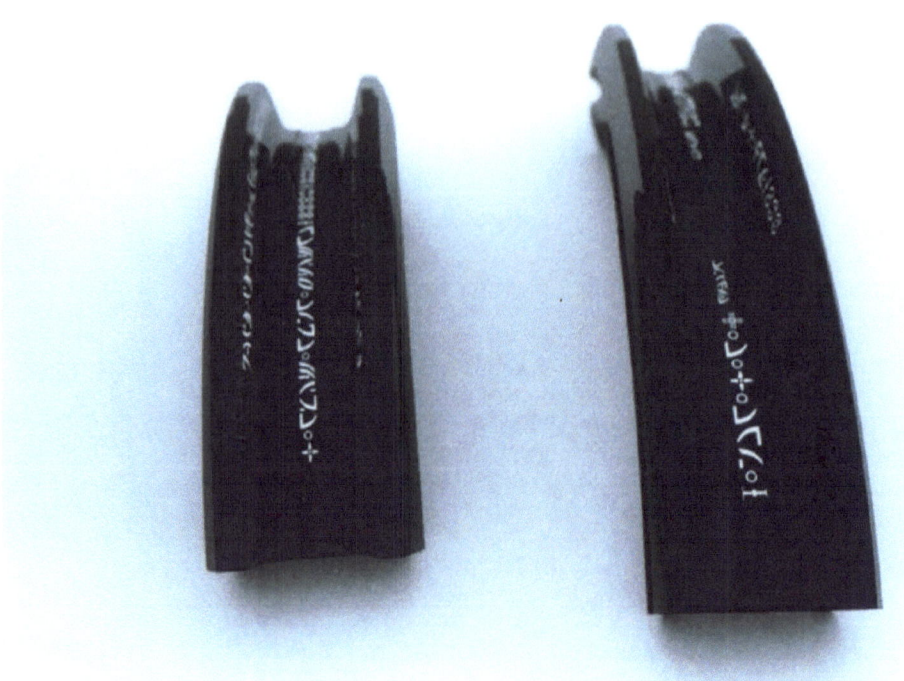

COSMIC MECHANICS

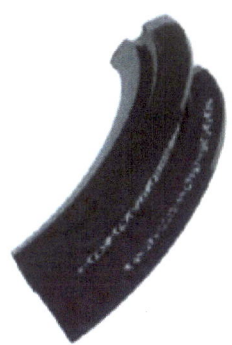

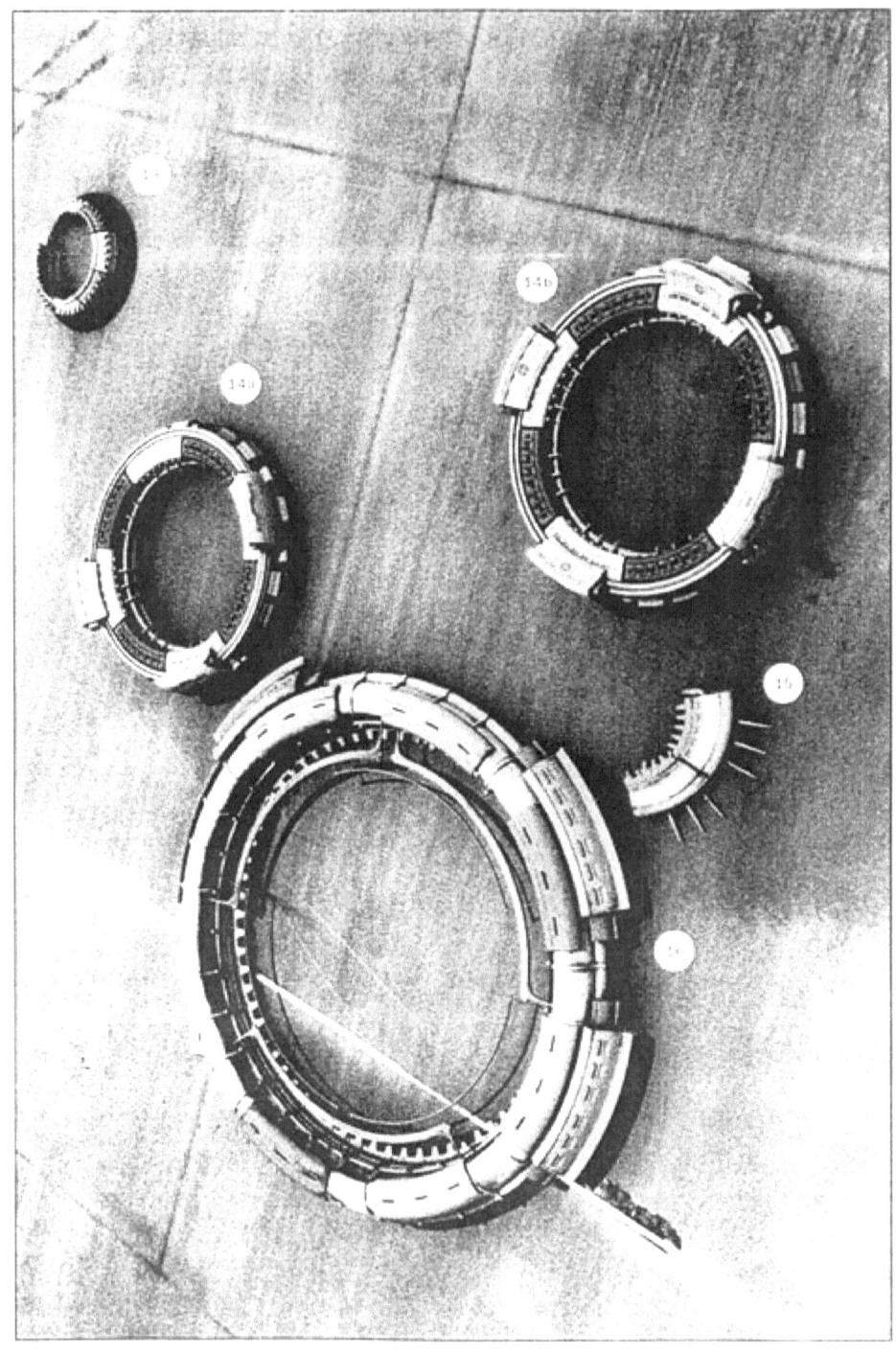

COSMIC MECHANICS

PALO ALTO CARET LABORATORY — LINGUISTIC ANALYSIS PRIMER

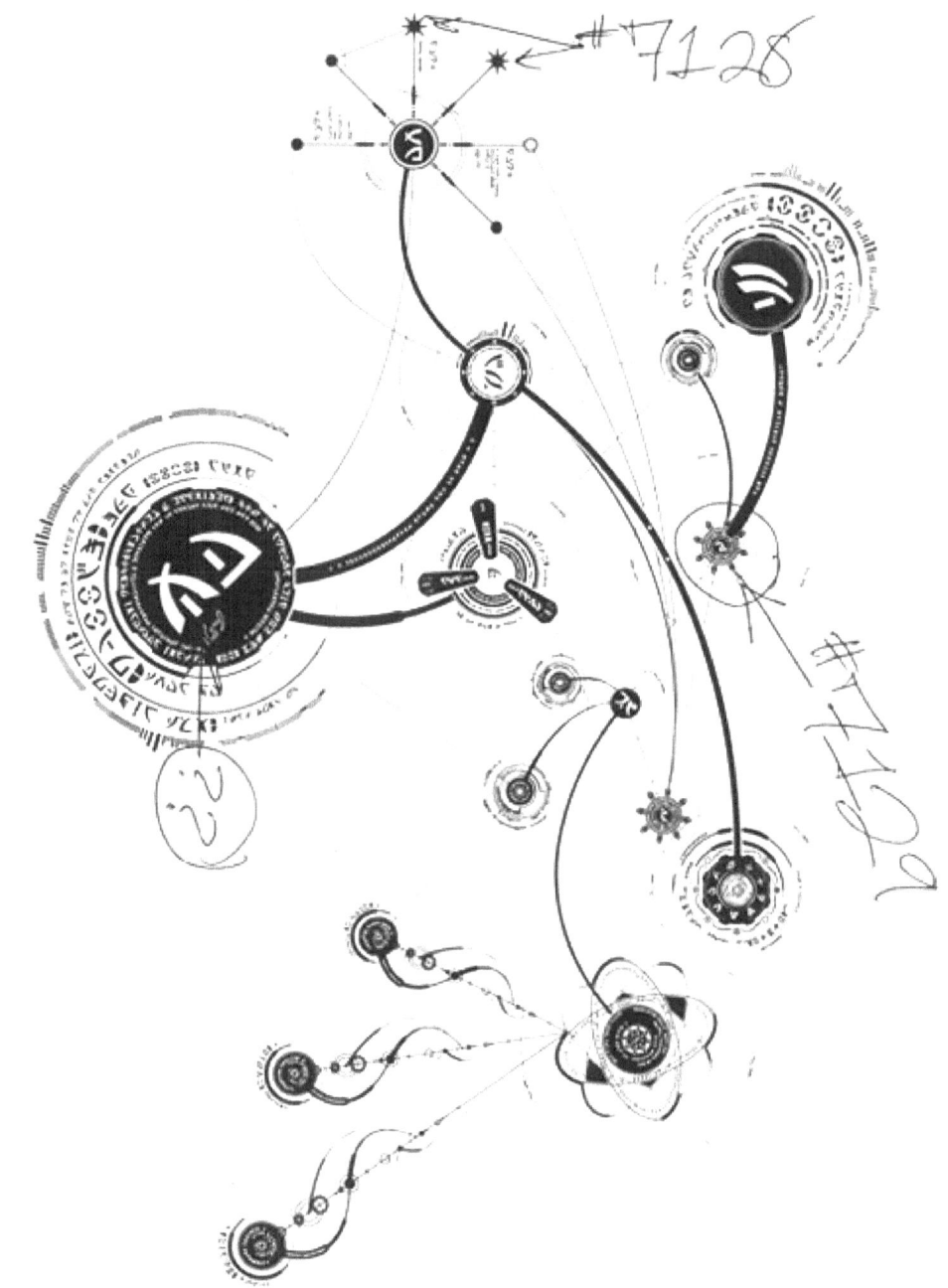

Figure 14.11
Full view of diagram D39-98-117.

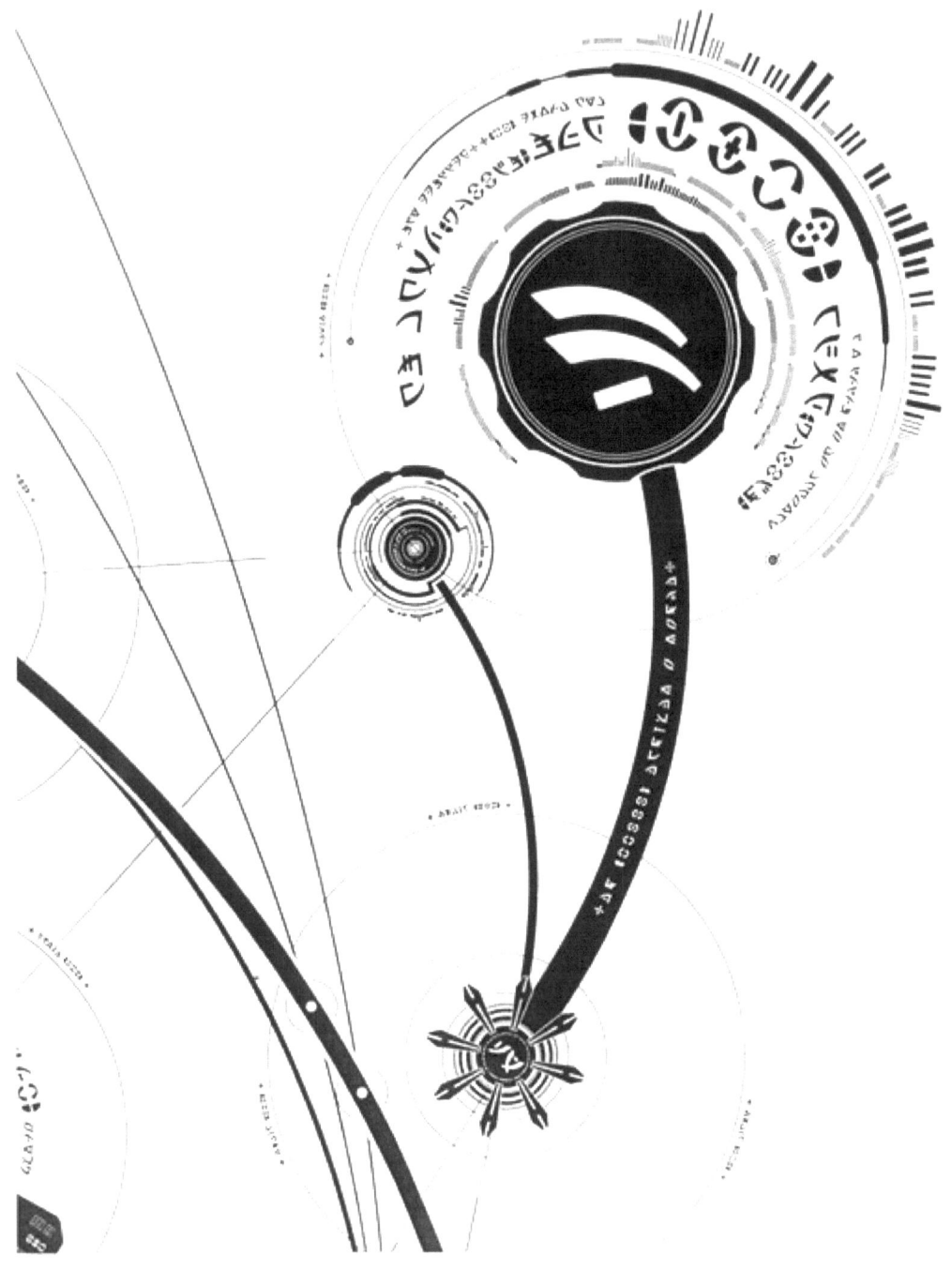

Figure 14.13
Rotary junction with orbital sub-junction connecting to an octal switch

Environmental Implications

We possess classified energy generation and antigravity propulsion systems capable of completely replacing all forms of currently used energy generation and transportation systems. These devices access the ambient electromagnetic and Zero Point Field to produce vast amounts of energy without the creation of pollution. These systems create energy by tapping into the ever present Zero Point Field, the baseline energy from which all energy and matter is fluxing. These are not perpetual motion machines and they do not violate the laws of thermodynamics. These systems do not require fuel to burn or atoms to split or fuse. They do not require central power plants, transmission towers or a multibillion dollar infrastructure to power remote areas. These systems can be set up anywhere energy is needed and they can be the solution to the major environmental problems of our world. The disclosure of these new technologies will give us a new sustainable civilization. World poverty will be eliminated and no place on earth will suffer from want. The deserts will bloom and there will be no limit to what man can achieve. But is this what our reptilian "custodians" desire for us? Does this fit in with their agenda? Will the classified technologies be released? No, not while they are holding us prisoners on earth and controlling us with our personal orbs. They need dumb, naive taxpayers who pose no threat to them. Our orbs are insidious and diabolical quantum computers which the reptilian "custodians" are using to ruin our lives. **Let's be responsible cosmic beings. Let's get rid of our orbs and go free!**

UFO's over Vancouver, B.C. 7 pm PST, 3 July, 2007 "We saw what looked like a glowing orb behind trees in the sky. It was moving slowly. About a half hour later people were looking up and asking me what are those? I looked up and lo and behold around 20 craft were directly above us! They hovered there for about 10 minutes. Looking at them through binoculars you could see beams of light emitting from them of different colors. They were moving slowly and going from side to side. It was witnessed by 30 or more people! We started snapping pictures as fast as we could. Then they left at high speed east. It was very exciting. This was in Maple Ridge, B.C, near Vancouver." – Rob

84) EXAMPLES OF SCALAR WAVES

Quantum waves are scalar. They are not like waves in the electromagnetic spectrum. They have an intensity, but no direction like electromagnetic waves have. They can be transmitted, manipulated, resonated and controlled like any other kind of wave. Energy can apparently appear out of nowhere with strange interference patterns describing grids. Here are some photos over Santa Barbara, California, of Soviet weather engineering utilizing scalar wave technology associated with "woodpecker signals" in the short wave radio bands.

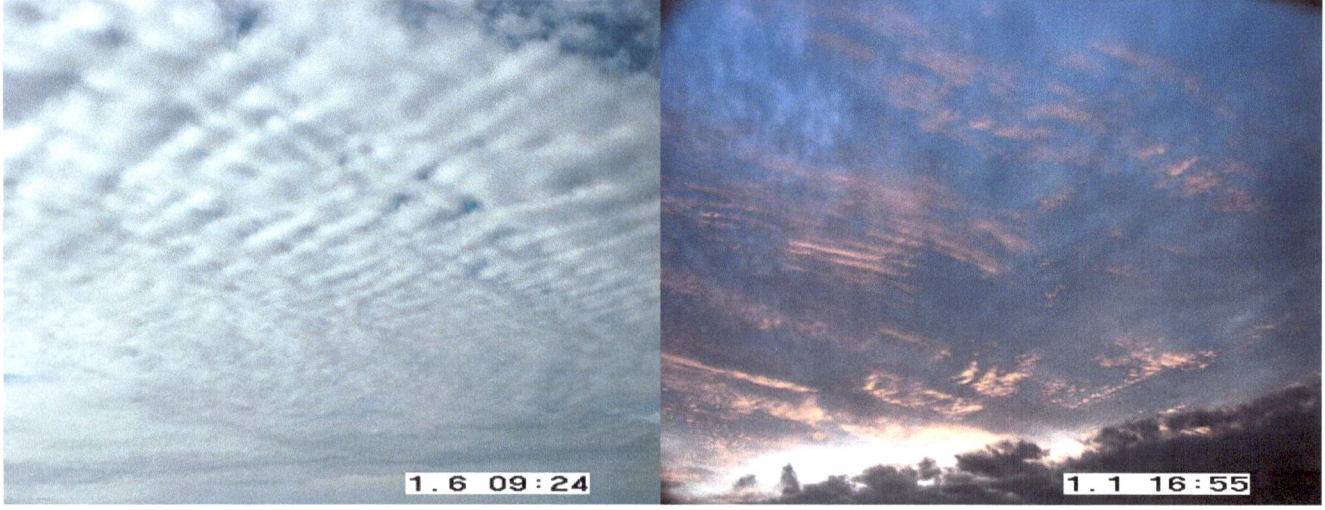

85) CONCLUSIONS

Now that I understand Dr. Wolff's discovery that there is no discrete particle or solid chunk of anything in the whole Physical Universe, that it is all standing waves which are communicating with each other, I finally understand the lights that I saw fly over the desert at Area 51. They were cosmic beings who joined up in two powerful groups to create those balls of light or kabalahs. Those beings were just like me before I was born into my present body. When my body dies I will fly away with my friends again. The Hessdalen lights are a similar group of cosmic beings. Were they on a mission to save our dying planet? Was it an attempt to control men's minds? Were they going to appear as UFO's over Las Vegas or create crop circles? I don't know what their purposes were but I do know how they operate. I know Cosmic Mechanics. **Here are the conclusions that I have reached in my knowledge quest:**

1. **I am Cause**. Prime Creator (God) is ALL THAT IS. Therefore, I am Prime Creator of my universe, expressing Itself as me. Since All That Is contains *everything*, I and everyone that is, has ever been, and ever will be, in all time, and in all parallel dimensions and realities must be contained within the All that is, Prime Creator, First Cause, God. This is the nature of existence.
2. **In the distant past I was practically a god and now no one is permitting me to be rehabilitated.**
3. **I take full responsibility as soon as I recognize that I am the cause of every effect I have ever experienced.**
4. **I have a viewpoint from which I create space.**
5. **I can create light and energy of any frequency and life itself.**
6. **I can duplicate perfectly any effect I have ever perceived.**
7. **My considerations define my reality**.
8. **I will never die**. I am a cosmic being. Bodies composed of wave structures will fall, I will persist.
9. **No one has ever allowed us to know how great we really are.** We were only given inhibitions.
10. **Matter is only solid because I consider it so**. Dr. Milo Wolff has proven that there is no discrete particle or solid chunk of anything in the whole Physical Universe, not even at the center of heavy atoms. It is all standing waves.
11. **If it is still, it is true**. If it is vibrating, it's illusion. And everything in this universe is vibrating.
12. **Waves in collision is the only communication there is**. There is no such thing as passive communication.
13. **My considerations in the past have been forcefully impressed in my orb.**
14. **My orb now distorts my reality since it has grown more powerful than me**.
15. **I'm getting rid of my orb, my reactive mind.**
16. **I am not striving for perfection as I am already perfect. What I am striving for is to *manifest* my perfection and to vanquish my 21 gram orb.**

17. **Everything in my reality is created by my beliefs and postulates.** As aspects of the All that is, Prime Creator God, I create my own personal reality as God creates, through thought. **Everything** begins with thought and postulates.
18. **As a Creator, I am 100% responsible for all creation.** It is not possible for anyone, or anything, other than me to be responsible for my experiences if I am the only one who can create in my reality. Every man is cause of his participation in the Physical Universe. I even take responsibility for other's cause.
19. **Everything happens for a reason.** Since God is Absolute Perfection, then nothing that God creates is pointless or extraneous. Therefore, as an individualized aspect of God, expressing Itself as me, the same applies to that which I create in my reality, whether or not I am consciously aware of it.
20. **There are no accidents--Nothing happens by chance.** Since everything that is created has a reason and a purpose, then the concept of an accident must be an illusion designed to avoid taking responsibility for the creation.
21. **I create 100% of the time. I have the power to choose whether I apply that 100% toward creating what I want or energizing what I don't want.** As an aspect of the Infinite Creator, I, too, am infinitely and constantly creating. How I choose to flow my thoughts and emotions determines what I attract into my world.
22. **All events are neutral. There are no negative experiences, although I can choose to *define* any experience as negative. Therefore, if I am experiencing negativity in my life, then I have chosen it.** Negative or positive definitions are a personal choice of a value judgment placed on any given situation or event, for the purpose of the experience. Since nothing can be in my reality unless I create it, then if it is there, and it is negative, then I have chosen for it to be there; and I have chosen to perceive it as negative. I always have another choice.
23. **I cannot be afraid unless I believe that whatever I fear has more power than I do.** But I am a matter, energy, space time production unit and nothing can possibly hurt me. Believing that anything/anyone else has more power than I do **IS** the erroneous belief that causes all fear.
24. **How I feel is a reaction to what I believe and postulate.** I make my own feelings so I'm going to indulge myself in enthusiasm and exhilaration! Writing this book is such a trip!
25. **There are no victims. There are only lessons.** If God is All that is, then there is only One of us here. Victim and victimizer are the same one idea. Whichever role I choose to be in

any situation is mirroring to me the false beliefs I hold inside, for the purpose of uncovering them. Left unrevealed, and unacknowledged, my erroneous beliefs will attract to me their energy equivalent in physical form. To alter the manifested effect of my error thinking, I only need to change my perception and release all judgment, which is the Cause of my creation that is being mirrored to me. All apparent victim/victimizer situations offer me the opportunity to overcome judgment and learn about my greatness.

26. **I am the only Authority in my life.** If I am the only one who can create in my reality, then I *must be* the only Authority of it. However, in an attempt to avoid taking responsibility for being that Authority, I can choose to falsely believe, that someone or something else has power over me.

27. **There are no limitations.** If I have an experience that I define as a limitation, it is happening as a result of the way I think. The way I think is controlled by my orb. I am overcoming my orb by disagreeing with it and getting the upper hand. I am also getting help from counselors. The *appearance* of all limitation is merely an effect of error thinking. God has no limitations; and as an aspect of God, neither do I, unless I **believe** or **postulate** that I do. In which case I will create the *appearance of limitation* as a direct result of my postulated beliefs.

28. **I have achieved my immortality** by conceiving my relationship to my body, my mind, and the Physical Universe. I will never die although my body won't last.

You are a cosmic being. Manifest the glory of God that is within you. It is not just in some of us; it is in everyone.

86) THE END F. Lee Aeilts, BSEE, LLD

1410 L. Ron Hubbard Way, Los Angeles, CA 90027 (323) 663-8047 Cell: (323) 373-6921

dr.aeilts@sbcglobal.net   www.cosmicmechanics.com

Other books by Dr. F. Lee Aeilts

GET RID OF YOUR DIABOLICAL ORB: It's an insidious demon

Extraterrestrial Artifacts And The Secret Of Human Identity

www.ingramcontent.com/pod-product-compliance
Lightning Source LLC
Chambersburg PA
CBHW051016180526
45172CB00002B/377